T0300326

Controlling Asbestos in Buildings

Asbestos dust is well-known for causing cancer and other life-threatening illnesses yet still contaminates countless schools, factories and office buildings. This raises the issue of the best way to deal with asbestos; immediate removal, containment or removal at renovation or demolition. Originally published in 1986, this report aims to evaluate these methods of dealing with asbestos in relation to their cost-effectiveness to conclude the most appropriate solution. This title will be of interest to students of Environmental Studies and Economics

Controlling Asbestos in Buildings

An Economic Investigation

Donald N. Dewees

First published in 1986
by Resources for the Future, Inc.

This edition first published in 2016 by Routledge
2 Park Square, Milton Park, Abingdon, Oxon, OX14 4RN
and by Routledge
711 Third Avenue, New York, NY 10017

Routledge is an imprint of the Taylor & Francis Group, an informa business

Publisher's Note
The publisher has gone to great lengths to ensure the quality of this reprint but
points out that some imperfections in the original copies may be apparent.

Disclaimer
The publisher has made every effort to trace copyright holders and welcomes
correspondence from those they have been unable to contact.

A Library of Congress record exists under LC control number: 86042611

ISBN 13: 978-1-138-96101-2 (hbk)
ISBN 13: 978-1-315-66004-2 (ebk)

CONTROLLING ASBESTOS IN BUILDINGS

AN ECONOMIC INVESTIGATION

DONALD N. DEWEES

A STUDY FROM **resources FOR THE FUTURE**

Washington, D.C.

Printed in the United States of America

Published by Resources for the Future, Inc.
1616 P Street, N. W., Washington, D.C. 20036

Books from Resources for the Future are distributed worldwide by The Johns Hopkins University Press

Library of Congress Cataloging-in-Publication Data

Dewees, Donald N.
Controlling asbestos in buildings.

Bibliography: p.
1. Asbestos in building—Safety measures.
2. Asbestos in building—Economic aspects.
3. Asbestosis—Mathematical models. I. Resources
for the Future. II. Title.
TA455.A6D49   1986   363.1'79   86-42611
ISBN 0-915707-27-6

The paper in this book meets the guidelines for permanence and durability of the Committee on Production Guidelines for Book Longevity of the Council on Library Resources.

Donald N. Dewees is Professor of Economics and Professor of Law at the University of Toronto. He wrote this study while at Resources for the Future as a Gilbert White Fellow.

CONTENTS

ACKNOWLEDGMENTS

This report was written while the author was a Gilbert White Fellow at Resources for the Future during 1984 and 1985. I would like to thank RFF and the Gilbert White Fellowship program for making this research possible. I would also like to acknowledge financial support from the Department of Economics of the University of Toronto for computer time, from a leave grant from the Social Sciences and Humanities Research Council of Canada, and from a grant from the Connaught Foundation to the Law and Economics Program at the Faculty of Law of the University of Toronto. The author was director of research for the Ontario Royal Commission on Asbestos from 1980 to 1984. The current work owes much to the understanding of asbestos problems developed while at the commission. Helpful suggestions on the development of the models were provided by Michael Toman and Alan Krupnick of RFF and Larry Smith of the University of Toronto. Clifford Russell, John Ahearne, Lester Lave, and two referees provided valuable comments on earlier versions of the manuscript.

Toronto, June 1986   Donald N. Dewees

one

INTRODUCTION

Of all the indoor pollutants, asbestos is probably the most studied because of the magnitude of the health effects that it causes for occupationally exposed workers. It is also most costly to control because of its wide use in sprayed and troweled insulation and in pipe and boiler insulation. The spray application of insulation-containing asbestos fibers was introduced in 1932 in Great Britain and soon spread to the United States and Canada. This process was originally used for thermal insulation, decorative purposes, and acoustical control, but after 1950 it was used primarily as a fireproofing to protect structural steel from heat in case of fire. Use of sprayed asbestos fireproofing grew through the 1960s and far exceeded the use of asbestos-containing sprayed material for thermal, decorative, and acoustical purposes. Such spraying essentially ended in 1973 partly in response to regulations in both Canada and the United States. The application of asbestos-containing insulation to pipes, boilers, and furnaces began in the 1920s, employing preformed thermal insulating sections or slabs, asbestos-cement compounds, and other materials. This use of asbestos continued until the early 1970s (RCA, 1984, chap. 9). It has been estimated that two-thirds of all asbestos used in the United States between 1890 and 1970 was utilized in the construction industry (Selikoff, 1980).

The workers who installed this asbestos in buildings were exposed to considerable concentrations of airborne asbestos fibers that in turn caused a well-chronicled occupational health disaster. In 1964 Dr. Irving Selikoff and others published the first of a series of studies that followed the health experience of members of the International Association of Heat and Frost Insulators and Asbestos Workers, who were primarily involved in installing asbestos insulation in buildings.

1

Between 1967 and 1976, about one-third of the deaths among these workers were attributable to their workplace asbestos exposure (Selikoff, Hammond, and Seidman, 1979). The continued study of these workers and of workers exposed to asbestos in other situations provides a tragic but uniquely valuable understanding of the relationship between exposure to a hazardous material and its health effects. It is this extensive study that allows the estimation of the dose-response relationship, and thus the prediction of the disease that may result from a given exposure to the substance.

The principal causes of premature death among those who have worked with asbestos are asbestosis, lung cancer, and mesothelioma, although other cancers have also been linked to asbestos exposure. Asbestosis is a chronic restrictive lung disease caused only by exposure to asbestos fibers. By reducing lung function, it can cause disability or death. While asbestosis has been a major cause of premature mortality among the asbestos workers in the Selikoff study, current exposures in buildings are to sufficiently low concentrations of asbestos that there is no risk that building occupants will develop asbestosis. Lung cancer may be caused by many factors, most notably smoking, but exposure to airborne asbestos fibers increases the risk of contracting this usually fatal disease. Mesothelioma is a rare cancer of the surface lining cells of the lung and abdomen. It occurs predominantly among those exposed to asbestos and, like lung cancer, is often fatal within a year of diagnosis. Thus the primary health risk posed by current exposures of building occupants and workers to airborne asbestos in buildings consists of an increased possibility of contracting lung cancer and mesothelioma.

The asbestos that causes these diseases is itself not a single material, but a family of hydrated silicates, of which the most important commercially are chrysotile, amosite, and crocidolite. These three types of asbestos differ in chemical composition and physical characteristics, including fiber shape. Chrysotile, which accounts for over 90 percent of asbestos used today, contains magnesium, is white or gray, consists of curly fibers, and is resistant to alkalis but not to acids. Amosite contains iron or magnesium or both, is generally brown, has straight needle-like fibers, has low tensile strength, and is moderately resistant to alkalis and acids. Crocidolite contains sodium and iron, is blue in color, has straight needle-like fibers, and has good acid resistance. Product uses that demand specific properties of the asbestos will usually require that a particular type of asbestos be used (table 1-1).

The focus of public policy concern regarding asbestos during the 1970s was worker exposures during the application of asbestos-con-

Table 1-1. Characteristics of Three Types of Asbestos

Characteristic	Chrysotile	Amosite	Crocidolite
Color	White-gray (green, yellow, pink)	Gray-yellow to dark brown	Blue
Chemistry	Hydrous silicate of Mg	Silicate of Fe and Mg	Silicate of Na and Fe plus water
Crystal structure	Fibrous and asbestiform	Prismatic, lamellar to fibrous	Fibrous
Texture	Fine, soft to harsh; silky	Harsh and coarse but pliable	Soft to harsh but flexible
Lustre	Silky	Vitreous-pearly	Silky-dull
Hardness	2.5–4.0	5.5–6.0	4.0
Resistance to acids	Attacked fairly rapidly	Attacked slowly	Good
Resistance to alkalies	Very good	Good	Good
Spinnability	Very good	Fair	Fair
Tensile strength (000 psi)	80–100	16–90	100–300
Decomposition temperature (C)	450–700	600–800	400–600
Major products	Textiles, flooring, cement products, friction material, "paper" products	Insulation products	Pressure pipes

Source: Ontario Royal Commission on Asbestos (1984) pp. 80–81.

taining friable insulation in buildings and the exposure of industrial workers to airborne asbestos dust. The installation of asbestos-containing friable insulation is now prohibited. Allowable exposures of workers to airborne asbestos fibers in the manufacture of asbestos-containing products have been greatly reduced. The use of asbestos in American industry appears to be declining, and the number of workers exposed is a small fraction of the number exposed in the past. Debate continues in the United States over the further reduction of allowable industrial exposures below the current limit of two fibers per cubic centimeter (f/cc),[1] and over an Environmental Protection

[1]See, for example, the notice of the Emergency Temporary Standard (ETS) in which the Occupational Safety and Health Administration (OSHA) proposed to reduce the permissible exposure limit for asbestos from 2.0 f/cc to 0.5 f/cc. *Federal Register* vol. 48, no. 215, Nov. 4, 1983, pp. 51086-51140. This ETS was subsequently stayed by a court challenge.

Agency (EPA) proposal to ban the production of a number of asbestos-containing products.[2] However, there has been no consensus that the need for action is urgent. In Ontario, Canada, the allowable exposure of workers to amosite and crocidolite asbestos is limited to 0.5 and 0.2 f/cc, respectively, and the allowable exposure of workers to chrysotile asbestos is limited to 1.0 f/cc.

While concerns about the high exposure of workers to asbestos in the *past* have been addressed, the problems posed by asbestos *now present* in existing buildings are more difficult to deal with. Many buildings contain asbestos, which will remain in place until it is removed, and many of these same buildings will still be standing a century from now. Removal, unfortunately, is neither costless nor riskless. It inevitably involves some asbestos exposure for the removal workers and, if poorly done, may cause risks for building occupants. Thus asbestos control work may reduce aggregate hazards, but at the same time shift them from one group to another. Building owners have obvious incentives to ignore an asbestos problem where possible, and to postpone expensive control actions for as long as possible. Building occupants and, in the case of schools, the parents of building occupants, may demand that any asbestos be removed at once. Parents, alarmed by the discovery of asbestos in their children's schools pursuant to an inspection and notification program mandated by the EPA's "Asbestos in Schools" rule,[3] have confronted school boards across the continent and demanded corrective action. Caught between alarmed parents and fiscal restraints, the boards find little guidance as to when expensive control actions are warranted and when they are unnecessary. Within the EPA there has been controversy over the degree of urgency with which asbestos in schools should be treated, as evidenced by a substantial softening of the tone of the agency's 1985 Guidance Document (EPA, 1985) in comparison with the 1983 Guidance Document (EPA, 1983).

The tension between occupants and owners is at the center of the current differences over what to do about asbestos currently in buildings. With respect to asbestos in buildings, the public policy debate is now focused on three issues. First, should any restrictions be placed on the installation of the remaining asbestos-containing prod-

---

[2]For discussion of the proposed regulations, and their blockage by the Office of Management and Budget, see "EPA Reverses Position on Asbestos Regulation," *Washington Post*, March 12, 1985, p. A15. The proposed regulation itself is published in 51 *Federal Register* 3738, January 29, 1986. On June 20, 1986, OSHA published a final rule setting an exposure limit of 0.2 f/cc for all types of asbestos. 51 *Federal Register* 22612, June 20, 1986.

[3]"Friable Asbestos-Containing Materials in Schools, Identification and Notification Rule," 47 *Federal Register* 23360, 40 *Code of Federal Regulations*, Part 763.

ucts, such as vinyl-asbestos floor tile and asbestos-containing roofing felts? While the U.S. EPA has proposed such restrictions, the concerns motivating these restrictions are centered as much on the workers manufacturing the products as on the release of asbestos from the products once in the building. Second, who should pay for controlling asbestos-containing material already in buildings? This cost has thus far fallen primarily upon the building owners, principally school districts and governments. Lawsuits have been launched by building owners against the installers, distributors, and manufacturers of these products to recover the cost of removing and replacing the material. Third, what should be done with asbestos-containing friable material already in buildings? Here the choices include removing the material at once, controlling it in some other way, or leaving it to be removed at renovation or demolition of the building. This study focuses on the third of these issues, and in particular on the economic implications of choosing different times to remove asbestos.

Curiously, the debate over these issues has included relatively little economic analysis, although it is clear that the potential cost of control programs is an important element in the reluctance to act precipitously. Few studies have calculated the cost-effectiveness of control programs.[4] This study begins to develop a methodology for economic analysis of asbestos control programs in buildings, and presents the results of three case studies. The goal is to illuminate in economic terms what is at stake when choosing among policies dealing with this problem.

The policy choices include ignoring the asbestos, removing it now, removing it at demolition, and several intermediate strategies. Regulatory agencies may encourage or compel one or more of these actions. Current regulations, however, clearly impose one constraint on the choice set, which the author fully supports: friable asbestos-containing material must be removed prior to demolition of the building. The question therefore is not *whether* to remove the material, but *when* to remove it. In this context, three issues are examined in this study. First, what is the effect on the economic life of a building of requiring that asbestos be removed before the building is demolished? Because the cost of removal effectively raises the cost of demolition, one might suspect that this would change an owner's calculation of the most economical time for demolition. Second, what is the cost-effectiveness of asbestos removal as a means of reducing health risks? More precisely, if the purpose of asbestos removal is to reduce the

---

[4]Two examples are Putnam, Hayes, and Bartlett (1984), and Research Triangle Institute. (1985)

risk of premature death, what is the cost of a particular program or policy per unit of risk reduction? Finally, what might be the financial impact on a building owner of discovering asbestos in his building, particularly if the owner will be required to pay compensation in the event of any deaths arising out of asbestos exposures in the building?

Since the issues addressed here are inherently empirical, the analysis must take place in the context of particular buildings. A set of three case studies is defined, each representing a particular building. It does not appear feasible to define "typical" buildings, so each case is presented not as a representative building of its type, but rather as an example. An exception is the second case, which is based on the average characteristics of public schools in the United States. The analysis is performed for each case study, and the sensitivity of the results to various of assumptions and parameters is explored.

The results of this analysis may be briefly summarized. The typical exposure of building occupants to airborne asbestos fibers is one one-thousandth to one ten-thousandth those of insulation workers in past decades. The risks of asbestos-related mortality from such exposures is on the order of sixteen fatalities per million persons exposed for ten years. This low health risk to building occupants when the asbestos is in good condition and undisturbed, combined with the high cost of removing asbestos insulation from a building, cause the cost of asbestos removal programs to be quite high when measured against the number of life-years saved. Most of the cases and variations studied here yield costs in excess of $1 million per life-year saved from early removal of the asbestos, far above the costs of other public programs that might reduce risks of mortality. As a means of improving public health, a crash program to remove all asbestos from public buildings would be very costly in relation to the modest health benefits achieved. Furthermore, asbestos removal workers may experience substantial exposures to asbestos, so that a program of early asbestos removal, possibly reducing aggregate health risks, may at the same time shift risks from occupants to removal workers. With respect to the effect of asbestos on the timing of building demolition, we conclude that a simple requirement that asbestos be removed prior to building demolition will tend to prolong the life of the building, but not by more than a few years in most cases. Finally, all conclusions from a study like this must be qualified by the uncertainties about our knowledge of the health effects of exposure to very low concentrations of airborne asbestos, by limited information about the cost of maintaining worker safety while the asbestos remains in a building, and by the variability of conditions from one building to another.

Now that the outlines of the study have been drawn, some caveats for the reader are in order. This study does not provide practical technical advice on what to do with asbestos in a particular building. The U.S. EPA (1983, 1985) has produced a series of useful reports and guidelines on this subject, as have various organizations concerned with asbestos in buildings. Nor does the study argue that economic analysis, whether through the determination of cost-effectiveness or otherwise, should be the sole basis for determining the appropriate response to environmental hazards. Although any analysis, including that performed here, must necessarily exclude factors not subject to quantification, these factors nevertheless deserve to be weighed in arriving at a decision. Finally, the cases presented here are examples, and do not exhaust the universe of relevant situations. The range of possible exposure levels, removal costs, exposures of removal workers, and other parameters is so great that they cannot all be captured in a small set of cases. Thus many actual situations may fall outside the range of parameters investigated here. The purpose of this analysis is to show how, in principle, one might examine the three issues we address and to present some sample results.

The author became interested in asbestos in buildings while serving as the Director of Research for the Ontario Royal Commission on Asbestos (RCA), which in May of 1984, after four years of work, presented to the Minister of Labour a three-volume, 900 page report. Chapters 9 and 10 of that report deal with asbestos in buildings, defining the problems, and making thirty-five formal recommendations. Subsequently as a member of the Advisory Council on Occupational Health and Occupational Safety, he participated in a review of the Ministry of Labour's "Proposed Regulation Respecting Asbestos on Construction Projects and in Buildings and Repair Operations,"[5] a regulation which incorporates many elements of the RCA's recommendations. This work for the commission forms the background for much of the work reported here.

This study should be of interest to those charged with developing policies for controlling asbestos in buildings, and to those who must make decisions for individual buildings, including school boards and building owners generally. It should also be of interest to those interested in developing economic analyses of the problem of asbestos in buildings. While an economic perspective permeates the study, noneconomists should find it intelligible, with the exception of chapter 5, which can be skipped without serious loss of continuity.

[5]Ontario Ministry of Labour, Occupational Health and Safety Division, November 16, 1984.

REFERENCES

EPA. *See* U.S. Environmental Protection Agency.

Ontario Royal Commission on Asbestos. 1984. *Report of the Royal Commission on Matters of Health and Safety Arising from the Use of Asbestos in Ontario* (Toronto, Queen's Printer).

Putnam, Hayes and Bartlett, Inc. 1984. "Cost and Effectiveness of Abatement and Asbestos in Schools." Draft (U.S. Environmental Protection Agency, Office of Pesticides and Toxic Substances, August 8).

Research Triangle Institute. 1985. *Asbestos Abatement Rules: A Preliminary Cost-Effectiveness Analysis.* Revised Draft Report to the U.S. EPA, Office of Toxic Substances, May.

RCA. *See* Ontario Royal Commission on Asbestos.

RTI. *See* Research Triangle Institute.

Selikoff, Irving J. 1980. "Asbestos-Associated Disease," in John M. Last, ed., *Public Health and Preventive Medicine*, 11th ed., chap. 13 (New York, Appleton-Century-Crofts).

Selikoff, Irving J., E. Cuyler Hammond, and Herbert Seidman. 1979. "Mortality Experience of Insulation Workers in the United States and Canada, 1943-1976," *Annals of the New York Academy of Sciences* vol. 330 (December 14) p. 103.

U.S. Environmental Protection Agency, Office of Toxic Substances. 1983. *Guidance for Controlling Friable Asbestos-Containing Materials in Buildings*, EPA 560/5-83-002 (Washington, D.C., U.S. Environmental Protection Agency, March).

_____. 1984. *Evaluation of the Asbestos-in-Schools Identification and Notification Rule*, EPA 56015-84-005 (Washington, D.C., U.S. Environmental Protection Agency, October).

_____. 1985. *Guidance for Controlling Asbestos-Containing Materials in Buildings*, 1985 ed. (Washington, D.C., U.S. Environmental Protection Agency).

two

ASBESTOS EXPOSURES IN BUILDINGS

The presence of certain asbestos-containing materials in buildings may raise the risk of exposure to elevated airborne asbestos fiber levels for building occupants and for building custodial, maintenance, renovation, or demolition workers. This section describes the materials that may raise such risks, the methods of measuring airborne asbestos fiber levels, and the exposures that may be experienced by building occupants and workers.

Asbestos Products Used in Buildings

Asbestos is a generic term referring to a family of minerals, of which the most important commercially are chrysotile, amosite, and crocidolite. Chrysotile is white or grey, and accounts for most of the asbestos used in construction in the past and today. Amosite or brown asbestos, and less commonly crocidolite or blue asbestos, were also used for building insulation products. As the next chapter of this report will show, there is continuing debate over the relative health hazards caused by breathing asbestos of these three types. The U.S. regulatory agencies, including the Occupational Safety and Health Administration (OSHA) and the Environmental Protection Agency (EPA) have treated all three types of asbestos as if they presented equal health risks. Regulatory agencies in the United Kingdom and Ontario have regulated chrysotile less strictly than the other two types of asbestos on the grounds that chrysotile is substantially less of a health hazard.

A large number of products used in the construction industry contain, or have in the past contained, asbestos (see RCA, 1984, chap. 9; and EPA, 1983, app. C for extensive discussions of asbestos-containing building products). When the asbestos-containing product is a liquid that would encapsulate the asbestos fibers after application, as with asbestos-containing paints, there is little likelihood that the fibers will become airborne. When the asbestos-containing product consists of a hard material encasing the asbestos, as with vinyl-asbestos floor tiles, it is unlikely that the fibers will become airborne except when the product is cut with power tools or sanded. But when the asbestos product is friable, or crumbly, fibers may be released when the material is installed, when it is disturbed by cleaning or maintenance work, or when it is contacted by a building occupant.

The most commonly found friable asbestos-containing products in buildings are sprayed insulation, troweled insulation, pipe and boiler insulation, and insulation board. Sprayed fireproofing was applied by either a wet or a dry process. In the wet process, chrysotile or amosite asbestos (at 5 to 30 percent by weight of the total formula) and some other fibers were mixed with a binder, such as cement or gypsum, and water and then sprayed onto the building surface. The resulting product is relatively hard and dense. The dry process was more common than the wet process and tended to use either chrysotile, usually at 5 to 70 percent by weight, or amosite, usually at 50 to 90 percent by weight, with some other fibers and binder. In North America, crocidolite was less commonly used. It was substituted, at 50 to 90 percent by weight, in a small number of dry process applications. The dry mixture was blown through a nozzle and wetted by a water spray as it left the nozzle. This dry process resulted in a material with much lower density in place and with greater thickness than the wet process material. Troweled insulation consisted of a mixture like the wet process mixture, but was applied by trowel rather than spraying. This resulted in a smoother surface and a more dense insulation. Troweled insulation was often used on exposed surfaces, giving a stucco effect.

Pipe and boiler insulation (or lagging) materials included preformed insulation sections, asbestos-cement compounds, corrugated asbestos paper, and some miscellaneous products. Preformed sections maintained their shape during shipping and installation, and could be cut to size upon installation. Asbestos usually represented 15 percent by weight of the product, and in any event not more than 80 percent. Corrugated asbestos paper consists of almost pure asbestos and may be made into preformed insulation sections for pipes. Irregular fittings such as elbows and valves were covered by asbestos cement, which consists primarily of short asbestos fibers and in some cases a binder of Portland cement, which is mixed with water and applied by hand to form a dense material when dried. These materials were used for thermal insulation on boilers and pipes carrying steam or hot water.

Asbestos insulation board is a friable fibrous board made of asbestos and other fibers and binder, and used for rough walls and ceiling tiles. This product does not release significant fibers in normal use, but considerable fiber levels may result from cutting the board. This should not be confused with asbestos-cement sheets which are very hard, not friable, and not easily damaged.

The use of asbestos in pipe and boiler insulation began in the 1920s and continued until the 1970s. The use of asbestos in sprayed and troweled insulation began in the 1930s, and grew rapidly in the 1950s when it was used for fireproofing as well as for thermal insulation (RCA, 1984, chap. 9). The spray application of insulation containing more than one percent asbestos ceased in North America about 1973, and is now prohibited in the United States and in Ontario by specific regulations. The application of moulded or wet-applied pipe and boiler insulation containing asbestos also ceased in about 1973 in North America, and is now prohibited in Ontario and the United States by regulations.[1]

Prevalence of Friable Asbestos Products in Buildings

Evidence on the prevalence of friable asbestos products in buildings in North America is sketchy, in part because there is no comprehensive building inventory to provide a population of buildings within which a sample could be inspected for asbestos. The most comprehensive studies of asbestos in buildings, discussed below, have pertained to public school buildings. Only one comprehensive national nonschool survey is known.

The U.S. EPA commissioned a telephone survey in the winter of 1983–84 of 1,800 public school districts and 800 private schools across the United States to determine, among other things, the prevalence of asbestos in schools. Asbestos-containing friable material (ACFM) was found in 35 percent of all schools surveyed (EPA, 1984a, pp. 5–17). In at least 45 percent of these schools (that is, in 45 percent of the school districts, many of which must include more than one

---

[1]See 40 Code of Federal Regulations 61, sec. 61.148(a) and sec. 61.150 for the prohibitions in the United States. See sec. 3 of "Regulation Respecting Asbestos on Construction Projects and in Buildings and Repair Operations," Ontario Regulation 654/85 for the prohibition in the Ontario regulation.

school) the ACFM was limited to pipe wrap (EPA, 1984a, pp. 5–20). Thus, less than 20 percent of all schools have sprayed or troweled ACFM (table 2-1). If the schools are analyzed by year of construction, the proportion of schools that contain ACFM is quite steady at about one-third for construction dates from 1900 to 1968, and dropping to 11 percent for the decade 1969–1978 (EPA, 1984a, pp. 5–22). In a separate study of schools containing ACFM, Constant (1983, p. 60) found that 56 percent of all sites in those schools contained ACFM. A site was defined as a student activity area, including a classroom, corridor, gymnasium, locker room, cafeteria, kitchen, library, or auditorium.

During the early 1980s, surveys were conducted in Ontario, Canada to determine the prevalence of ACFM in educational institutions. These surveys found that 20 percent of all public school buildings contained ACFM, and that 9 percent contained exposed sprayed asbestos (RCA, 1984, p. 597). It thus appears that applying sprayed asbestos-containing insulation to exposed surfaces was somewhat more common in the United States than in Ontario.

The U.S. EPA commissioned another sample survey, conducted during 1983 and 1984, to determine the proportion of buildings in the United States that contain ACFM. A total of 231 buildings were inspected in ten cities selected to be representative of the continental United States, but excluding schools and state and local government buildings. Estimates were prepared separately for private nonresidential buildings, federal government buildings, and rental apartment buildings with more than 10 units (table 2-1). The survey found that 20 percent of all buildings had some ACFM, with 5 percent containing sprayed or troweled ACFM, and 16 percent containing pipe and boiler ACFM (EPA, 1984b, pp. 2–3). ACFM was least prev-

Table 2-1. Prevalence of Asbestos in U.S. Buildings

Building type	Percentage of buildings with ACFM		Percentage of area with sprayed or troweled ACFM
	All types	Sprayed/troweled	
Schools[a]	35	20	
Federal[b]	39	16	10
Residential[b]	59	18	7
Nonresidential[b]	16	4	1.8
All buildings[b]	20	5	3

[a] EPA (1984a), p. xii. Of school buildings with ACFM, 56% of *sites* have ACFM (Constant, 1983, p. 60).
[b] EPA (1984b), pp. 2-3, 7-9, 7-12.

alent in nonresidential buildings, 16 percent of which contained ACFM and only 4 percent of which contained sprayed or troweled ACFM. Surprisingly, 18 percent of residential buildings contained sprayed or troweled ACFM, although the asbestos content at 9 percent was relatively low. This ACFM appears to represent primarily decorative or acoustic troweled material in apartment buildings. These residential buildings also contained the highest proportion of pipe and boiler insulation (44 percent) of any building type.

Several other patterns appear in the data. The last column in table 2-1 shows that the proportion of the *area* of buildings with sprayed or troweled ACFM is always less than the proportion of all *buildings* with sprayed or troweled ACFM. This occurs because many buildings containing ACFM do not contain that material throughout the building. Asbestos-containing ceiling tiles were found only rarely, and then the asbestos content was low, 3 percent or less. Only a small proportion of the buildings in the survey had sprayed asbestos concealed behind drop ceilings; most of the sprayed asbestos was completely exposed (EPA, 1984b, pp. 7–16). With respect to year of construction, a much higher proportion of the buildings constructed during the 1960s had sprayed or troweled ACFM than buildings constructed at other times, a result consistent with the observation that spraying of asbestos insulation grew rapidly starting in the late 1950s. Pipe and boiler insulation, in contrast, is equally common in buildings dating back well before World War II.

The data discussed above consistently show that pipe and boiler ACFM is much more common than sprayed or troweled ACFM. Asbestos in pipe and boiler insulation is often easy to control by wrapping the insulation with cloth or other material that will contain the insulation and prevent the release of asbestos fibers. The control of asbestos in sprayed or troweled insulation is considerably more difficult, since wrapping is not usually feasible. Furthermore, pipe and boiler insulation containing asbestos is often confined to areas of a building not frequented by the public. Thus the actual problem of controlling ACFM in buildings is less daunting than would be the case if all ACFM in buildings consisted of sprayed or troweled insulation.

Measurement Methods

The concentration of asbestos fibers in the air is measured by drawing an air sample through a filter and counting the asbestos fibers captured on the filter. An optical microscope is used to count fibers when the air sample is expected to contain large concentrations of asbestos fibers, but is inappropriate for analyzing ambient air or the air in buildings where asbestos fiber concentrations are generally quite low and where other fibers may be numerous (Chatfield, 1983, p. 18). The transmission electron microscope (TEM) is the preferred instrument for analyzing the low asbestos fiber concentrations of building or ambient air. The filter is prepared for the TEM by either the direct or indirect method. Direct preparation involves fixing fibers to the filter on which they were originally caught. This has the advantage that the fibers may be observed with a minimum of disturbance. A disadvantage is that rough handling of the filter before fixing may dislodge and loosen some fibers. Indirect preparation involves ashing the filter, dispersing in water the indestructible material, which is mostly asbestos, followed by refiltering the residue and then direct preparation analysis. Indirect preparation allows a better view of all fibers, but may break up fibers and fiber bundles, thus raising the fiber count above that which would have resulted from observing the original undispersed sample.

The TEM analysis involves identifying each fiber on a filter segment, and noting the dimensions and fiber type of each asbestos fiber. The mass of each fiber may be calculated from its size and the known density of that type of asbestos. The results may therefore be reported as a number of fibers or as a mass of fiber material. The volume of the air sample drawn through the filter is known, so it is possible to report the results of the filter analysis as the number of asbestos fibers or the mass of fibers per volume of air.

There is considerable support for the proposition that the hazards presented by airborne asbestos fibers are much greater for fibers longer than 5 or perhaps 8 micrometers, than for shorter fibers (RCA, 1984, p. 261). Furthermore, since about 1970, workplace asbestos fiber measurements have reported only fibers longer than 5 micrometers. Evaluating the health risks posed by airborne asbestos fiber concentrations in buildings is therefore facilitated if the building fiber count includes only those fibers longer than 5 micrometers, which are easily segregated if direct preparation is used with the TEM. However, since the indirect preparation method separates some fibers, most studies using this analytical method report only the mass of the asbestos fibers.

There is, as yet, no international agreement on the appropriate method for measuring asbestos fiber concentrations in building air. Furthermore, the direct and indirect preparation methods appear to yield significantly different fiber concentrations, at least with respect to fibers shorter than 5 micrometers, although for longer fibers, the results are quite similar (Chatfield, 1984). Thus the use of data sets from different researchers poses problems for the development of consistent estimates of airborne asbestos fiber levels.

Standardization of Reported Fiber Concentrations

Epidemiological studies that relate asbestos exposure to human health generally express the asbestos exposure in terms of the airborne fiber concentration in fibers per cubic centimeter (f/cc) measured by the optical microscope, but rarely in terms of mass measurement. Measurements of asbestos exposure in buildings are generally made by TEM, and expressed either in terms of mass (nanograms per cubic meter, ng/m^3) or f/cc. Predicting health effects from building exposures therefore requires some conversion so that those exposures may be expressed as a number of fibers longer than 5 micrometers that would be seen by an optical microscope, per cubic centimeter.

The asbestos fibers in an air sample will be of varying lengths and diameters. The optical microscope can detect fibers longer than 5 micrometers, and with diameters greater than about 0.4 micrometers. The TEM can detect fibers far shorter and thinner than the smallest that are visible with an optical microscope. Even if the TEM is used to count only fibers longer than 5 micrometers, it will generally detect several times as many as an optical microscope. If the TEM fiber count reports only total fibers, the count will be far greater than a count only of fibers longer than 5 micrometers. When the TEM analysis is reported in terms of fiber mass, the number of fibers longer than 5 micrometers will depend upon the size distribution of the fibers in the particular air sample. The size distribution of fibers varies greatly from one situation to another. Since the various analysis and reporting methods are based on different portions of this size distribution, no single conversion factor can accurately relate the reporting methods to each other. Any conversion factor must therefore be recognized as an average only, and subject to considerable uncertainty.

Despite these formidable problems, conversion factors have been determined to provide a basis for evaluating airborne asbestos levels in buildings. The Ontario RCA (1984, p. 574) concluded that thirty optically measured fibers longer than 5 micrometers would be associated with one nanogram measured by TEM, which implies that 1.0 f/cc measured optically equals 33,000 ng/m^3 measured by TEM.

In doing so, it adopted the EPA conclusion that thirty optically measured fibers longer than 5 micrometers represent one nanogram of mass measured by TEM (EPA, 1980, p. 76). The RCA (1984, p. 569) further concluded that a TEM count of fibers longer than 5 micrometers should be divided by ten to determine the number of optically visible fibers. The National Academy of Sciences (NAS) (1984, p. 88) emphasized the uncertainty in conversion, but concluded that one f/cc measured optically would equal sixty total fibers measured by TEM, and 30,000 ng/m³. This implies that an optical microscope sees a smaller fraction of all fibers than the RCA believes, but the mass of fibers in one f/cc is very similar to that found by the RCA. This report will adopt the assumptions of the Ontario RCA that one f/cc measured optically equals 33,000 ng/m³ measured by TEM.

Exposure of Building Occupants

The sampling of building air, and the analysis of the resulting filters using the TEM, costs $500 or more per sample where a sample is a single filter exposed in a single location. A thorough analysis of a building might involve sampling of the air in a number of rooms, at a cost of $500 per sample. This high cost has severely limited the study of airborne asbestos levels in buildings. Table 2-2 presents the results of four major studies of airborne asbestos levels in buildings in which sprayed or troweled asbestos insulation was present. All studies report the asbestos concentration in mass terms. The Pinchin

Table 2-2. Asbestos Fiber Exposures in Buildings with Sprayed Asbestos

Study	Mass concentration (ng/m³)	Optically visible[a] (f/cc)
Sebastien, et al. (1980)		
Outdoors		
arithmetic mean	0.96	0.00003
99% less than	7.0	0.0002
7 asbestos-free buildings		
arithmetic mean	1.9	0.00006
21 buildings with sprayed asbestos		
median	5	0.00015
highest building average	70	0.0021
94% of measurements less than	100	0.003
highest single reading	751	0.023

Table 2-2. Asbestos Fiber Exposures in Buildings with Sprayed Asbestos (Continued)

Study	Mass concentration (ng/m³)	Optically visible[a] (f/cc)
Nicholson, Rohl, and Weisman (1975)		
Outdoors		
83% less than	20	0.0006
96% less than	50	0.0015
highest	87	0.0026
2 asbestos-free buildings (12 samples)		
92% less than	20	0.0006
highest	42	0.0013
17 buildings with sprayed asbestos		
wet-applied (28 samples)		
93% less than	20	0.0006
highest	180	0.0054
dry-applied (53 samples)		
53% less than	20	0.0006
92% less than	100	0.003
highest	830	0.025
Pinchin (1982)[b]		
19 buildings (63 samples)		
median building	0.13	0.0
90% less than	8.0	0.0002
highest building average	11.0	0.0003
Constant, et al. (1983)[c]		
Outdoors (25 schools)		
median	0.9	0.00003
maximum	40.6	0.0012
Indoor controls (19 schools)		
median	21.8	0.0007
maximum	362.0	0.011
Indoor sites (48 sites)		
lowest	0.002	6×10^{-8}
median site	92.7	0.003
90% less than	422.0	0.013
maximum	644.0	0.02

[a] Except in the case of the Pinchin study, the optically visible fiber count is computed as ng/m³/33,000.
[b] The Pinchin study provides a count of fibers longer than 5 micrometers. The fiber concentrations for the three conditions presented in the table are: 0.0 f/cc; 0.002 f/cc; and 0.003 f/cc. The optically visible fiber count presented in the table is calculated as 1/10 of the TEM count of fibers longer than 5 micrometers. Pinchin (1982), p. 6.7.
[c] Constant, et al. (1983), p. 60.

study also reports fiber counts. The last column presents an estimate of the concentration of optically visible fibers implied by the mass concentration or, in the Pinchin study, implied by the TEM count of fibers longer than 5 micrometers. The analytical methods differ among the studies, so even the mass measurements are not necessarily compatible.

What is the best way to summarize a set of measured airborne asbestos fiber levels for purposes of estimating the average exposure experienced by persons in the room, building, or set of buildings? The distribution of airborne asbestos fiber measurements is skewed, with many low readings, and a few very high readings. If the logarithm of these measurements is plotted, *it* is approximately normally distributed, which means that the measurements are approximately log-normally distributed. The true exposure for a situation in which the sample measurements are log-normally distributed is better presented by the *geometric mean*, which is lower than the arithmetic mean or average, but will be close to the median.[2] When summarizing sets of measurements of airborne asbestos levels, this study will, where possible, use the geometric mean or the median, rather than the arithmetic mean.

The data in table 2-2 show considerable variability in indoor airborne asbestos levels within each study, and still greater variations among the studies. Pinchin (1982) and Constant (1983) found little or no correlation between measured fiber levels and any of the hazard indices designed to indicate the risk of excess exposure to asbestos dust. Sebastien (1980) and Constant (1983) found a clear progression from low exposures outdoors to moderate exposures indoors in buildings or rooms without asbestos, to higher exposures in buildings or rooms with sprayed asbestos. Nicholson (1975) did not find a great difference between outdoor levels, indoor levels in asbestos-free buildings, and indoor levels in building with sprayed asbestos. Pinchin (1982) found indoor levels so low that they may not be significantly above outdoor levels.

The levels reported by Pinchin are well below those of the other

---

[2]The arithmetic mean, or average, is the sum of the sample measurements divided by the number of measurements. The median is the measurement such that half of the measurements are less than the median, and half are more than the median. The geometric mean is the antilog of the average of the logs of the measurements. When the sample measurements are distributed log-normally, the arithmetic mean is highest, and the median is close to the geometric mean. The best estimate of the average exposure U given a set of sample readings x is obtained by taking the log of each sample reading. Let the log of x be y. The mean of the sample logs is Y, and the variance of the y is s^2. The best estimate of the true average exposure U is: $U = \exp.(Y + 1/2s^2)$. See Greenwood and Hartley (1962) p. 412.

three studies. This may result from the use of the direct preparation method in the Pinchin study in contrast to the indirect preparation method in the other studies, or from differences in the fiber identification methods used in the studies. The analysis of Pinchin's sample employed a very rigorous methodology for determining whether observed fibers were actually asbestos. Alternatively the differences in levels may arise from different conditions in the asbestos itself in Ontario and in the other studies. Most of the asbestos in the Pinchin study was not exposed, while most of the asbestos in the U.S. schools studied by Constant was exposed and accessible. More recent measurements taken by Pinchin and analyzed by the direct preparation method have continued to yield low exposures.[3]

The Constant study of twenty-five schools in a large urban school district in the United States is the most recent of the studies reported in table 2-2. It reports airborne asbestos concentrations considerably above those of the other three studies. The median concentration in a room *without* ACFM, in a building in which other rooms have ACFM is 21.8 ng/m^3, which is twice the highest building average found by Pinchin, and four times the median building concentration found by Sebastien. The median concentration in a room with ACFM is 92.7 ng/m^3, which is in the top 10 percent of the readings of any of the other studies. The high readings found by Constant are as puzzling as the low readings found by Pinchin. The U.S. EPA, however, has apparently had second thoughts about the weight to be given to the Constant study. In its 1985 guidance document the EPA refers to the Constant study and to a subsequent, but as yet unpublished, study by Chesson as the basis for a conclusion that airborne asbestos concentrations in schools range from about 1 ng/m^3 to 100 ng/m^3 (EPA, 1985, pp. 1–4). This is a reduction of one to two orders of magnitude from the 80 to 800 ng/m^3 attributed to the Constant study in the 1983 guidance document (EPA, 1983, pp. 1–5). At the same time, the tone of the guidance document has become less alarming regarding the risks faced by building occupants. This suggests that when the Constant and Chesson studies are considered together, they are consistent with the other evidence that airborne concentrations of asbestos fibers in buildings are generally well below 100 ng/m^3.

The Constant study shows that rooms in asbestos-containing buildings that do not themselves contain asbestos experience considerably lower airborne asbestos levels than do rooms that contain asbestos. Thus the average exposure in an asbestos-containing building is usu-

[3]Personal communication from D. J. Pinchin, August 21, 1985.

ally significantly lower than the exposure in an asbestos-containing room in that building.

The data in table 2-2, together with the 1985 EPA guidance document, support the conclusion that most exposures in buildings containing sprayed asbestos insulation will be the equivalent of less than the 0.001 f/cc measured optically. The Constant data suggest that average exposures in U.S. schools containing sprayed or troweled asbestos insulation might be the equivalent of 0.002 f/cc measured optically, but the subsequent Chesson study seems to refute the suggestion that school exposures are this high. The Nicholson and Sebastien data suggest that the majority of exposures in buildings with sprayed or troweled asbestos insulation will be the equivalent of less than 0.001 f/cc measured optically. There is, however, considerable variability in indoor asbestos concentrations, and some rooms and some buildings may experience asbestos levels that are well above or well below these averages. All of these conclusions, furthermore, are subject to some uncertainty, and one cannot be confident that the differences in the averages just reached are not artifacts of the particular sampling or analytical methodology employed. In contrast to these low exposures, the insulation workers, who are suffering from a terrible toll of disease, experienced average exposure levels over their working lives of between 3 f/cc and 15 f/cc (RCA, 1984, pp. 551–557). The current exposures of building occupants therefore range from one one-thousandth to one ten-thousandth of the exposure intensity which the insulation workers experienced.

The studies reviewed above do not separately analyze the airborne asbestos concentrations caused by pipe and boiler insulation. In many of the buildings that contain only pipe and boiler insulation, that insulation is confined to boiler rooms and other limited portions of the building. Pipe and boiler insulation is relatively easy to enclose with wrappings of cloth or metal cladding. It seems fair to conclude that pipe and boiler insulation that is in good condition and properly wrapped should not cause significant elevations of indoor airborne asbestos fiber levels unless it is disturbed. The conclusions in the preceding paragraph regarding occupant exposures apply to buildings with sprayed or troweled insulation and overstate the exposure of occupants of buildings in which the only ACFM consists of pipe and boiler insulation in good condition.

Exposure of Building Workers

While the exposure of passive building occupants to airborne asbestos fibers is generally very low, workers who disturb asbestos insulation may be exposed to asbestos fiber levels that exceed current workplace exposure limits in the United States and Ontario. Removal of asbestos insulation can cause very high fiber levels, as can spray painting to encapsulate the asbestos. Renovation or maintenance of the building can cause high fiber levels if the insulation is disturbed. Work above a suspended ceiling that encloses sprayed asbestos can result in elevated airborne asbestos fiber levels, which can cause risks for the workers themselves and for building occupants, if proper precautions are not observed.

While the large number of asbestos removal projects conducted during the last decade has required massive sampling of the exposure of workers, few studies of these exposures have been published. Table 2-3 reports the results of two such studies. Paik, Walcott, and Brogan (1983) studied the exposure of workers during asbestos removal at three buildings. In one building, the sprayed asbestos was saturated with water before and during removal to minimize airborne dust. When such wet methods were used, the geometric mean airborne fiber level was 0.5 f/cc. Pinchin (1982) found fiber levels with a geometric mean of approximately 0.35 f/cc in his study of four asbestos removal projects. When respirators are worn the actual exposure of the workers would be considerably less than the measured fiber levels. Dust levels are far greater when dry removal methods are employed. Paik, Walcott, and Brogan found fiber concentrations

Table 2-3. Airborne Asbestos Fiber Levels During Building Maintenance, Renovation, and Asbestos Removal

Activity	Number of observations	Fiber level Geometric mean (f/cc)	Maximum (f/cc)
Renovation[a]			
Carpenters	105	0.13	>2
Electricians	35	0.13	>2
Sheet-metal workers	37	0.19	>2
Painters	7	0.08	0.3
Removal of sprayed insulation			
Wet removal[b]	15	0.5	2–10
Wet removal[c]	58	0.35[d]	10.9
Dry removal[b]	79	16.4	>100

[a] Paik, Walcott and Brogan (1983), p. 431, table V.
[b] Paik, Walcott and Brogan (1983), p. 431, table VI.
[c] Pinchin (1982), p. 7.17.
[d] The figure presented here is estimated from Pinchin's data on arithmetic means, and some description of the distribution of individual readings.

21

with a geometric mean of 16.4 f/cc during removal work where it was not possible to use water on the asbestos. It is reported that dry removal of amosite routinely yields fiber concentrations exceeding 100 f/cc (Pinchin, 1986). Pinchin found that removal of pipe and boiler insulation could cause airborne fiber concentrations reaching 10 f/cc, but that improved work practices reduced this concentration to about 5 f/cc and that experience and continued vigilance further reduced the concentrations to about 1.0 f/cc. Pinchin concluded that care and experience were important determinants of fiber levels caused during the removal of pipe and boiler insulation.

Paik and coauthors also studied the exposure of workers to asbestos fibers during the renovation of eleven multistory buildings containing friable asbestos insulation. Table 2-3 shows the geometric mean of the exposures faced by four trades involved in the renovation work. In all cases, the geometric mean exposure was below 0.2 f/cc, less than the fiber levels experienced during wet removal. The highest exposures, with a mean of 0.19 f/cc, were experienced by sheet metal workers whose work caused the greatest disturbance of the insulation material. Pinchin (1982, pp. 7.5 and 7.8) found similarly low breathing zone fiber levels for workers carrying out inspections above a false ceiling that covered sprayed asbestos insulation, but considerably greater exposures for one worker performing maintenance work above such a ceiling. These measured exposures for renovation work in buildings containing ACFM are less than the occupational exposures currently allowed in the United States and in Ontario, except for work above a suspended ceiling.

Renovation and maintenance work that disturbs ACFM in a building, the disposal of asbestos waste, or the demolition of a building containing asbestos may cause elevated levels of airborne asbestos fibers for those in the vicinity of the building or of the disposal work. One purpose of the regulations adopted in Ontario and in the United States is to control this escape of asbestos dust from the work site so that airborne asbestos fiber levels outside the site are not significantly elevated. Regulations in Ontario and in the United States govern the disposal method for waste ACFM removed from a building, with the intended result of avoiding significant releases of asbestos fibers during the transportation and disposal of the asbestos waste.[4]

I conclude from these data that building maintenance or renovation that involves working around sprayed asbestos insulation will cause

[4]General-Waste Management Regulation, Ontario Regulation 175/83; 40 Code of Federal Regulations 61 Subpart M, secs. 61.152 and 61.156.

airborne fiber levels for the workers themselves that may range from 0.1 to 0.5 f/cc. Removal of sprayed asbestos insulation using wet methods and considerable care may cause airborne fiber levels of 0.3 f/cc or more. Dry removal of sprayed or of pipe and boiler insulation may cause airborne fiber levels exceeding 10 f/cc and sometimes 100 f/cc.

REFERENCES

Chatfield, Eric J. 1983. "Measurement of Asbestos Fibre Concentrations in Ambient Atmospheres," *Royal Commission on Asbestos Study Series*, no. 10 (Toronto, Ontario Royal Commission on Asbestos).

_____. 1984. "Measurement and Interpretation of Asbestos Fibre Concentrations in Ambient Air." Paper presented to the 5th AIA Colloquium, Johannesburg, South Africa, October.

Constant, Paul C. Jr., Fred J. Bergman, Gaylord R. Atkinson, and coauthors. 1983. *Airborne Asbestos Levels in Schools*, EPA 560/5-83-003 (Washington, D.C., U.S. Environmental Protection Agency, Office of Toxic Substances).

EPA. *See* U.S. Environmental Protection Agency.

Greenwood, J. A., and H. O. Hartley. 1962. *Guide to Tables in Mathematical Statistics* (Princeton, N.J., Princeton University Press).

NAS. *See* National Academy of Sciences.

National Academy of Sciences, Committee on Nonoccupational Health Risks of Asbestiform Fibers of the National Research Council. 1984. *Asbestiform Fibers: Nonoccupational Health Risks* (Washington, D.C., National Academy Press).

Nicholson, William J., Arthur N. Rohl, and Irving Weisman. 1975. *Asbestos Contamination of the Air in Public Buildings*, EPA 450/3-76-004 (Washington, D.C., U.S. Environmental Protection Agency, October).

Ontario Royal Commission on Asbestos. 1984. *Report of the Royal Commission on Matters of Health and Safety Arising from the Use of Asbestos in Ontario* (Toronto, Queen's Printer).

Paik, Nam Won, Richard J. Walcott, and Patricia A. Brogan. 1983. "Worker Exposure to Asbestos During Removal of Sprayed Material and Renovation Activity in Buildings Containing Sprayed Material," *American Industrial Hygiene Association Journal* vol. 44, no. 6 (June) pp. 428–432.

Pinchin, Donald J. 1982. "Asbestos in Buildings," *Royal Commission on Asbestos Study Series*, no. 8 (Toronto, Ontario Royal Commission on Asbestos).

_____. 1986. Personal communication.

RCA. *See* Ontario Royal Commission on Asbestos.

Sawyer, Robert N. and Charles M. Spooner. 1978. *Sprayed Asbestos-Containing Materials in Buildings: A Guidance Document, Part 2*, EPA 450/2-78-014 (Washington, D.C., U.S. Environmental Protection Agency, March).

Sebastien, Patrick, M. A. Billion-Galland, G. Dufour, and J. Bignon. 1980. *Measurement of Asbestos Pollution Inside Buildings Sprayed with Asbestos.* Translation of a document prepared for the Government of France, Ministry of Health and Ministry for the Quality of Life Environment, 1977, EPA 560/13/80-026 (Washington, D.C., U.S. Environmental Protection Agency, August).

U.S. Environmental Protection Agency, Office of Toxic Substances. 1983. *Guidance for Controlling Friable Asbestos-Containing Materials in Buildings*, EPA 56015/5/83-002 (Washington, D.C., U.S. Environmental Protection Agency, March).

_____. 1984a. *Evaluation of the Asbestos-in-Schools Identification and Notification Rule*, EPA 560/5-84-005 (Washington, D.C., U.S. Environmental Protection Agency, October).

_____. 1984b. *Asbestos in Buildings: A National Survey of Asbestos-Containing Friable Materials*, EPA 560/5-84-006 (Washington, D.C., U.S. Environmental Protection Agency, October).

_____. 1985. *Guidance for Controlling Asbestos-Containing Materials in Buildings*, EPA 560/5-85-024 (Washington, D.C., U.S. Environmental Protection Agency, June).

three

MODELS FOR
PREDICTING
ASBESTOS DISEASE

While a number of diseases have been associated with exposure to asbestos fibers, three diseases stand out as the principal causes of premature death among those who have worked with asbestos: asbestosis, lung cancer, and mesothelioma. In the past, asbestosis has been a major cause of death among asbestos-exposed workers. However worker exposure intensities today are a small fraction of those which caused the disease that is now occurring. It is generally believed that there is a safe threshold exposure level below which asbestosis will not occur, while for the cancers no such threshold has been shown to exist. One estimate places this threshold at a cumulative exposure of 25 f/cc-years; that is, twenty-five years of exposure to 1.0 f/cc (RCA, 1984, p. 281). It is likely that at current workplace exposures, the amount of asbestosis that will be caused will be zero or insignificant (U.K. Advisory Committee on Asbestos [UKAC], 1979, p. 58; RCA, 1984, p. 281). Exposures in buildings and in the ambient environment are as little as one one-thousandth of current workplace exposures, so there is no risk that building occupants or the general public outdoors will develop asbestosis. This disease may be ignored in projecting the health risks from exposing new workers to current fiber levels, or from exposing persons to prevailing asbestos concentrations in buildings or outdoors.

This leaves lung cancer and mesothelioma as the primary diseases of interest. Lung cancer has many causes, but it is now accepted that the inhalation of asbestos fibers in sufficient concentrations can increase the risk of this usually fatal disease. Mesothelioma is a rather rare cancer occurring in the surface cells lining the chest or abdominal cavity. Like lung cancer, mesothelioma is usually fatal within a year or two of diagnosis. Most studies now assume that the dose-response function for these cancers is a straight line through the origin, so there is no absolutely safe exposure level; any nonzero exposure causes a nonzero risk.

Gastrointestinal cancers and other cancers have been associated with asbestos exposure, but their incidence rates are considerably lower than that for lung cancer. OSHA (1983b, p. 51129) assumed that gastrointestinal and other cancers would cause mortality rates equal to 10 percent of the lung cancer mortality rate. However, the RCA (1984, p. 102) said that "the association between asbestos exposure and both gastrointestinal cancer and cancer of the larynx is neither as strongly nor as consistently established in the medical literature as the association between asbestos exposure and [lung cancer and mesothelioma]." We may ignore these cancers in estimating future disease rates without substantially underestimating total disease rates.

The problem of predicting the health effects of current exposures to asbestos is therefore reduced to the problem of predicting the incidence of lung cancer and mesothelioma that will result from the exposure of a cohort of specified characteristics to the anticipated concentration of asbestos fibers for a specified period of time. The problem studied here is fundamentally different from the problem of predicting aggregate future asbestos-related disease in North America arising from *past* asbestos exposure. One example of the latter type of study is that of Walker and coauthors (1983) who use mesothelioma as an indicator of asbestos exposure. They project future mesothelioma incidence and multiply this incidence by fixed ratios of the incidence of asbestosis and lung cancer to the incidence of mesothelioma. Other projections of future disease from past exposure are those of Peto, Henderson, and Pike (1981) and Nicholson and coauthors (1981). These studies also use methods that are appropriate for aggregate analysis, but omit the detail that can be incorporated in a study of a particular cohort exposed to a single asbestos type.

This chapter reviews the health effects models, used in several recent publications, that present quantitative risk assessments of the lung cancer and mesothelioma incidence likely to result from asbestos exposure. The sponsors of the studies are the Ontario Royal Commission on Asbestos (RCA), the Chronic Hazard Advisory Panel (CHAP) of the Consumer Product Safety Commission, the U.S. Occupational Safety and Health Administration (OSHA), and the U.S. National Academy of Sciences (NAS). The general principles behind the models are presented, followed by a detailed description of the specific model used by the Ontario Royal Commission on Asbestos. The remaining models are compared with the RCA model. The coefficients estimated for these models are presented and differences among them analyzed. The health risks predicted by each model for the exposure of a typical building occupant and for a typical workplace exposure are presented. The reader not interested in technical details of the disease models may skip the mathematical portions of this chapter without loss of continuity.

Lung Cancer Models

General Considerations

The lung cancer models assume that there is a background lung cancer rate existing without asbestos exposure that varies with the age and gender of the individual and his smoking habit. Exposure to asbestos *multiplies* the background lung cancer rate by an amount proportional to the extent of the asbestos exposure (Hammond, Selikoff, and Seidman, 1979). Since smokers have a much higher lung cancer rate than nonsmokers, an asbestos exposure which, for example, doubles the lung cancer rate will cause a much greater absolute increase in the number of lung cancer cases among smokers than among a similar population of nonsmokers. The background lung cancer rate is extremely low until almost age forty and rises thereafter. The maximum risk is for individuals in their late fifties and older.

The exposure of the individual to asbestos may be described by the average intensity of that exposure, measured in fibers per cubic centimeter (f/cc), and by the duration of the exposure. The models assume .that the cumulative exposure, defined as the product of intensity times duration, is a satisfactory measure of the dose and that the response, the probability of disease, is linearly related to this dose. While this assumption is reasonable for the intensities considered here, it may not be reasonable for very high intensities. In fact, the Ontario RCA concluded that short (several months), intense (tens of f/cc) bursts of exposure might overwhelm the body's defense mechanisms and cause a disproportionate risk of disease (RCA, 1984, chap. 5, p. 306).

The models assume that the excess risk of lung cancer is proportional to the cumulative exposure to asbestos as just defined. Epidemiological data are not now, and probably never will be, sufficient to prove that a linear relationship fits better than a nonlinear one. However the linear relationship is plausible biologically (RCA, 1984, chap. 5, E 3b), is easy to work with, and is conservative in that at low doses it predicts more disease than the competing models. All recent risk assessments have used the linear model.

Lung cancer is often referred to as a disease in which there is a latent period between exposure and disease. It might be argued that the observed delay is an artifact of the worker's first exposure usually occurring years before the age at which lung cancer can occur, as demonstrated by the age at which the background lung cancer risk becomes significant. Existing data are consistent with both the age relationship and with a latent period. Most of the models assume a ten-year delay between first exposure to asbestos and any elevation of the disease rate, although this assumption is not implemented in the same way in all models that use it.

In these models, the elevation of risk will continue until death, from whatever cause. This elevation of risk even in old age is particularly uncertain because few cohorts have been followed for a sufficiently long time after exposure ceased to test whether the risk continues or declines. One cohort that was studied did show an apparent trail-off of risk after thirty years after the first exposure (Nicholson, 1981). However, in the absence of robust data defining this phenomenon, the models generally assume no trail-off.

The Ontario Royal Commission on Asbestos Model

The RCA model calculates lung cancer mortality rates for a cohort at five-year intervals, then sums the mortality in each quinquennium to determine lifetime mortality for the entire cohort. The risk of lung cancer at any age applies only to those who have survived to that age. Mortality tables are used to determine the proportion of a population that will survive to any age, determined separately for males and females, for smokers and for nonsmokers. When these tables are used in the model, they are modified to allow for the deaths that occur as a result of asbestos-related diseases.

The RCA lung cancer model determines the lung cancer mortality that will occur in a population of N workers that is first exposed to asbestos at age A. The exposure in scenario i is of intensity $F(i)$, and continues for D years. No asbestos-related deaths are calculated until after the tenth year after first exposure, to allow for the latency associated with lung cancer. The annual number of excess deaths due to lung cancer attributable to asbestos exposure over the lifetime of this cohort, up to age eighty years, is

$$M_L(i,t) = K_L F(i) D N I_E(a) S(a)/S(A) \qquad (1)$$

where K_L = a proportionality constant derived from epidemiological studies which relates the cumulative dose of asbestos exposure to the excess relative risk of lung cancer

$F(i)$ = average intensity of asbestos exposure, scenario i

D = duration of exposure to asbestos years

N = number of workers exposed

A = age at first exposure to asbestos

a = age in year t, equal to $A + t$

I_E = the age-, gender-, and smoking-specific lung cancer mortality rate

$S(a), S(A)$ = the cumulative probability of survival to ages (a) and (A), respectively

The total number of asbestos-related lung cancer deaths over the life of the cohort is the sum of the mortality in each quinquennium from age A to age eighty. The mortality *rate* is the total mortality divided by N.

Other Lung Cancer Models

The other lung cancer models are quite similar in form to that used by the RCA. All multiply a gender- and smoking-specific lung cancer rate by a quantity including the cumulative exposure to asbestos. There are, however, some differences. With respect to latency, the RCA model ignores calculated deaths until the eleventh year after first exposure. It is not clear how the CHAP model deals with latency, since the text states: ". . . it is necessary to take into account a lag time of about 10 years for the effect of a given exposure to be manifest" (CHAP, 1983, p. 91). This may imply treatment similar to that of OSHA and EPA whose models incorporate the exposure variable lagged by ten years. The NAS model does not deal explicitly with latency. Doll and Peto (1985, p. 35) assume that a given exposure cannot contribute to the risk of lung cancer before five years have passed.

Another difference is the degree of aggregation of the models. The RCA model calculates mortality in five-year increments for the life of the cohort. The CHAP, OSHA, and EPA models also appear designed to determine mortality as a function of time since first exposure. The NAS model, however, is less useful for analyzing age- and time-related questions, since it performs a single calculation of the lifetime mortality of the cohort. The main equations of the models are presented in table 3-1, in a consistent format.

Table 3-1. Equations for Lung Cancer Incidence

Report	Equation[a]		Latency treatment
RCA[b]	$I_L(i,t) = K_L F(i) D I_E(a)$ $\qquad = I_E$	$t>10$ $t\leq10$	Omit mortality during first 10 years after first exposure
CHAP[c]	$I_L(i,t) = K_L F(i) D I_E(a)$		"Take account of a lag time of about 10 years"
OSHA[d]	$I_L(i,t) = K_L F(i) D(t-10) I_E(a)$ $\qquad = I_E$	$t>10$ $t\leq10$	Lag exposure 10 years
NAS[e]	$I_L(i,t) = K_L F(i) D I_E(a)$		No explicit latency

[a] All equations are gender- and smoking-specific. Definitions: I_L = predicted lung cancer incidence; i = an exposure scenario; t = years since first exposure; a = age; D = duration of exposure; $F(i)$ = intensity of exposure in scenario i; K_L = estimated dose-response coefficient for lung cancer; I_E = expected lung cancer rate absent asbestos exposure.
[b] 1984, p. 458.
[c] 1983, p. II-91.
[d] 1983a, p. 11.
[e] 1984, p. 208.

Mesothelioma Models

General

In contrast to lung cancer, mesothelioma is a disease related almost exclusively to asbestos exposure. There is, for all practical purposes, no background rate, and smoking does not appear to interact multiplicatively with asbestos to increase the mesothelioma rate as it does with lung cancer. It is therefore necessary to derive a model different from the lung cancer model that determines mesothelioma mortality as a function of the intensity and duration of the asbestos exposure, and the timing of that exposure.

The models reviewed here all are derived from the work of Julian Peto (1983). Peto finds that mesothelioma mortality depends not upon age, as does lung cancer, but upon the number of years since the first exposure to asbestos. In this model, the risk of mesothelioma mortality resulting from a quantum of exposure, for example exposure for one day to an intensity of F, will increase over time in proportion to the gth power of time since first exposure (t), where g may be determined empirically. As with lung cancer, there is a question whether there is a time delay between exposure and disease. The RCA model incorporates no delay for mesothelioma. If exposure is continuous from first exposure to death, the risk of death at any time t after first exposure is the sum of the risks created by each day of exposure which is proportional to the integral of time since first exposure. The integral of t^g is $t^{g+1}(g+1)$. If exposure is continuous from first exposure for a period of D years, the risk t years after first exposure where $t>D$ is the integral of the risks created by each day of exposure from 0 to D and is proportional to:

$$R = t^{(g+1)} - (t - D)^{g+1} \tag{2}$$

In the case of the lung cancer model, the relationship between the cumulative exposure to asbestos fibers and the lung cancer risk is linear through the origin. In the case of the mesothelioma model, there is also a linear relationship, but this time it is not between the cumulative exposure and the risk, but between the intensity of exposure and the disease risk. The duration of exposure is accounted for in the power function described above.

There is a question whether the mesothelioma risk continues to increase as indicated in equation (2) long after the onset of exposure. As in the case of lung cancer, the available data do not allow the determination whether the relationship described above continues until death, or whether there is a trail-off in risk. As in the case of lung cancer, the models have not introduced any trail-off because of the absence of data that suggest a functional form for that trail-off. It is entirely possible that such a trail-off does occur, in which case these models will overstate the risk of mesothelioma mortality.

The Ontario Royal Commission on Asbestos Mesothelioma Model

The RCA mesothelioma model, like the lung cancer model, calculates mortality rates for a cohort in five-year increments over the life of the cohort. The model determines the mesothelioma mortality that will occur in a population of N workers that is first exposed to asbestos at age A. The number of mesothelioma deaths attributable to asbestos exposure for a five-year period is

$$M_M(i,t) = 5 K_M (t^4 - (t - D)^4)F(i)NS(a)/S(A) \tag{3}$$

where K_M = a constant derived from epidemiological studies which relates the risk of mesothelioma mortality to the intensity and duration of asbestos exposure

    a = age

    A = age at first exposure

t = time since first exposure, = $a - A$
D = duration of exposure
$F(i)$ = intensity of exposure in scenario i
N = number of persons exposed
S = survival probability to the specified age

The factor of 5 appears in equation (3) because K_M is the *annual* risk, but the equation covers a five-year period. The total number of asbestos-related mesothelioma deaths over the life of the cohort is the sum of the mortality for all five-year age groups up to eighty years of age.

The risk of contracting mesothelioma predicted by equation (3) increases as the fourth power of the time since first exposure. This means that exposure to asbestos at a younger age can greatly increase the risk of disease. Thus, as will be shown later, mesothelioma becomes a particular concern when assessing the exposure of children to asbestos.

Other Models

The other models are very similar to the RCA model, with minor differences. Where the RCA model and Doll and Peto (1985, p. 33) use an exponent of 4 in equation (3), CHAP and OSHA use an exponent of 3, and the NAS uses an exponent of 3.2. The RCA and NAS models incorporate no delay before disease may occur, while CHAP, OSHA, and EPA all subtract ten years from t and D. The combination of the larger exponent and absence of a ten-year time delay in the RCA model cause this model to be more sensitive to the age of first exposure and to the assumed lifetime of the cohort than are the other models. The RCA model will therefore tend to predict more disease for a cohort first exposed at an early age, such as school children, than will the CHAP model, after estimating both models on the same workplace data. Which of these two models best reflects the empirical data is a close question, given the uncertainty in those data themselves. In particular, the range of uncertainty surrounding the value of the exponent in equation (3) is very great, which implies greater uncertainty in the predictions of mesothelioma risks for young children than in similar predictions for workers. As with lung cancer, the NAS model computes a single mortality figure for the lifetime of the cohort, rather than computing mortality by age groups within the cohort. The NAS model does not differentiate between the period of exposure and subsequent periods. While this is satisfactory for computing risks from exposures that begin at birth and are continuous until death, this model cannot be used for predicting risks in a case in which there is a finite period of exposure during an individual's life. The equations for the various models are presented in table 3-2.

Estimated Coefficients

Methodology

The lung cancer and mesothelioma models described above can be fitted to historical epidemiological data from about a dozen studies of the health experience of asbestos-exposed workers. These studies have gathered data on the exposure of workers and their disease experience including a count of deaths from various asbestos-related diseases. Each study focuses on workers in a particular plant, firm, or industry. All are historical prospective studies which began long after most of the workers had been first employed. The investigators defined a study cohort of workers, traced the work history of the cohort back in time, gathered data on the mortality experience of the cohort in the past, and then followed the cohort for the duration of the study. For a discussion of the methodology of the seven studies examined by the Ontario RCA and the quality of the resulting estimates, see the report of the RCA (1984, app. to chap. 7, sec. 2). The

Table 3-2. Equations for Mesothelioma Incidence

Report	Equation	Latency treatment
RCA[a] (also Doll and Peto)	$I_M(i,t) = K_M F(i)[t^4 - (t-D)^4]$	$t > D$
	$= K_M F(i)[t^4]$	$t \leq D$
CHAP[b] (also OSHA, EPA)	$I_M(i,t) = K_M F(i)[(t-10)^3 - (t-10-D)^3]$	$t > 10 + D$
	$= K_M F(i)(t-10)^3$	$10 + D \geq t > 10$
	$= 0$	$10 > t$
NAS[c]	$I_M(i,t) = K_M F(i)t^{3.2}$	None

Note: Definitions: I_M = mesothelioma incidence; K_M = mesothelioma dose-response coefficient; $F(i)$ = intensity of asbestos exposure in scenario i; D = duration of exposure; t = years since first exposure.
[a] 1984, p. 470.
[b] 1983, p. 92.
[c] 1984, p. 209.

studies relied upon by CHAP, OSHA, and NAS are discussed in their respective reports.

The data from each study may be used to derive the coefficients required for equations (1) and (3). In the case of lung cancer, the coefficient represents the quotient of the excess relative risk of lung cancer divided by the cumulative exposure. Each cohort can be divided into subcohorts, each with its own average age, cumulative exposure, and relative risk experience. Where data permit, regression analysis can be used to estimate the coefficient directly from a set of data points each of which represents a subset of the cohort. This involves estimating K_L in the equation for relative risk (RR)

$$RR = 1 + K_L FD \tag{4}$$

The underlying lung cancer rate and the survival rates shown in equation (1) need not be included in the regression because they are implicit in the notion of relative risk, in that they determine the underlying lung cancer risk against which the experience of this cohort is being compared. Where data are insufficient for regression analysis, the cohort is used to provide a single point estimate of the coefficient.

The mesothelioma coefficient K_M may be estimated similarly by fitting mortality data to the equations in table 3-2.

Evaluation of Estimates

Tables 3-3 and 3-4 contain the dose-response coefficients derived by the RCA, CHAP, OSHA, and NAS. The EPA uses the OSHA coefficients. The RCA analyzed seven studies, CHAP and OSHA analyzed eleven, and the NAS analyzed nine, only seven of which are included in these tables. Two of the NAS studies are excluded because no other survey analyzed them. OSHA presents not only the best estimate of the coefficient, but a "range of uncertainty" around the coefficients, shown in parentheses in the tables. These ranges of uncertainty are not precisely confidence intervals, but incorporate both the statistical strength of the estimated coefficient and other information about possible errors in input data. Two features of the coefficients are striking: the similarity among estimates of the coefficients for a given epidemiological study, and the differences between the coefficients from one epidemiological study to another.

The coefficients are quite similar for a given study, given the uncertainties in the underlying data. The CHAP and OSHA estimates differ by more than a factor of two only in the case of the Rubino

Table 3-3. Dose-Response Coefficients for Lung Cancer (K_L)

Epidemiological study	RCA[a]	CHAP[b]	OSHA[c]	NAS[d]
Mining (chrysotile)				
McDonald, et al. (1980)	0.00046	0.0006	0.00065 (0.0002–0.0011)	0.0006
Nicholson, et al. (1979)	—	0.0012	0.0023 (0.001–0.007)	0.0015
Rubino, et al. (1979)	—	0.0017	0.0051 (0–0.009)	—
Manufacturing (chrysotile) General				
Newhouse & Berry (1983)	0.00058	0.0006	0.0006 (0–0.008)	—
Textiles				
Peto (1980)	0.01	0.01	0.0076 (0.0009–0.023)	0.0007–0.008
Dement (1982)	0.0416	0.023 0.044	0.042 (0.023–0.21)	0.053
Manufacturing (mixed) Asbestos-cement (crocidolite)				
Finkelstein (1983)	0.042	0.048	0.067 (0.033–0.13)	—
Weill (1979)	—	0.0031	0.0033 (0.0016–0.0086)	—
Miscellaneous				
Henderson & Enterline (1979)	0.0007	0.0033 0.005	0.0047 (0.0068–0.0173)	0.003
Insulation (mixed)				
Selikoff et al. (1979)	0.0101	0.01	0.02 (0.008–0.030)	0.017
Seidman et al. (1979)	—	0.068	0.068 (0.0049–0.14)	0.091
Average	N.A.	0.003 – 0.03	0.01	0.02

Notes: Ranges of uncertainty are in parentheses. N.A. = not applicable.
[a] p. 491.
[b] p. II-129.
[c] p. 47a.
[d] p. 215.

study of miners. The CHAP and OSHA estimates differ from those of the RCA by more than a factor of two only in the case of the Henderson and Enterline study, in which case the RCA appears to have assumed that the historical exposure was greater by a factor of seven than the exposure assumed by the other analysts. In the case

Table 3-4. Dose-Response Coefficients for Mesothelioma (K_M)

Epidemiological study	$K_M \times 10^{-9}$			
	RCA[a]	CHAP[b]	OSHA[c]	NAS[d]
Manufacturing (chrysotile)				
Textiles				
Peto (1980)	7.2	7.0	7.0	8.5
			(3–20)	
Manufacturing (mixed)				
Asbestos-cement (crocidolite)				
Finkelstein (1983)	213	120	120	—
			(40–300)	
Insulation (mixed)				
Selikoff et al. (1979)	13.3	15	15	13.9
			(5–25)	
Seidman et al. (1979)	—	57	57	72.2
			(30–110)	
Average	N.A.	3–30	10	25.3

Notes: Ranges of uncertainty are in parentheses. N.A. = not applicable.
[a] p. 493.
[b] p. II-129.
[c] p. 47a.
[d] p. 217.

of the Newhouse and Berry, Peto, and Selikoff studies, the RCA and the CHAP coefficients are virtually identical. Thus, except for the Rubino and Henderson and Enterline studies, the four sets of estimated coefficients are reasonably consistent.

Doll and Peto (1985, pp. 34–39) present coefficient estimates of $K_L = 0.01$ and $K_M = 6.2 \times 10^{-9}$ for exposure to chrysotile in textile manufacturing. These coefficients are similar to those presented by the RCA and CHAP for the Peto study. Doll and Peto conclude that crocidolite and amosite are more hazardous than chrysotile, but they decline to estimate separate dose-response relationships for exposures to these types of asbestos.

The dramatic differences in coefficients are not between RCA, CHAP, OSHA and NAS, but rather between the various epidemiological studies themselves. The lung cancer coefficients based on the Dement study of textile workers and on the Finkelstein study of crocidolite-exposed asbestos-cement workers are over eighty times the coefficients based on McDonald's study of chrysotile miners and Newhouse and Berry's study of friction products workers. The Peto, Finkelstein, and Selikoff studies yield mesothelioma risks at least as great in absolute terms as the lung cancer risks, while the other studies find small or insignificant mesothelioma risks.

What should be done with the disparate risks suggested by the various epidemiological studies? We could assume that all are measuring the same risk and combine them. This is the approach taken by OSHA. After noting that "The range of estimates of risk from the eleven epidemiologic studies is rather large" (OSHA, 1983b, p. 51124) and that the range of values for the lung cancer coefficient alone covers two orders of magnitude, OSHA calculates the arithmetic mean of the lung cancer coefficients as 0.0201, and the geometric mean as 0.007. OSHA then concludes that "Considering the industrial processes other than mining and milling, OSHA believes 0.01 to be a reasonable estimate of [the lung cancer coefficient]" (OSHA, 1983b, p. 51125). The NAS also accepts averaging, calculating a median K_L as 0.011, which is then "rounded upward" to 0.02 (NAS, 1984, p. 214). The median K_M is calculated as 25.3 $\times$ 10^{-9}. CHAP finds reasons why averaging is not appropriate, but calculates a range within which the coefficient should lie. These averages are shown in the bottom line of tables 3-3 and 3-4.

There are good reasons, however, to believe that these studies are measuring risks that are different for different fiber types and different industrial processes. The three commercial types of asbestos differ considerably in their physical and chemical properties, as was shown in chapter 1. These differing physical and chemical properties have led to different product uses calling for different types of asbestos. Furthermore, industrial processes can change some properties of the asbestos fibers, and can mix them with other substances. Rather than expecting that all of these asbestos types and processes should lead to the same health risk, one should expect them to lead to different risks. This is in fact what is shown by the data recording the health experience of workers exposed to asbestos.

The RCA differentiates risks by fiber type, concluding that historical exposures to crocidolite and amosite were associated with higher disease rates than exposure to chrysotile. The RCA further concludes that mesothelioma is generally associated with exposure to amosite or crocidolite, and rarely with exposure to chrysotile. Finally the RCA concludes that for a single fiber type such as chrysotile, the health risk depends upon the fiber release process. Thus a given measured fiber concentration in mining and in brake manufacturing causes far less disease than the same measured fiber level in textile spinning and weaving. Working with sprayed insulation and pipe and boiler insulation causes intermediate risk levels (RCA, 1984, chap. 5).

The conclusion that the health risks from asbestos exposure depend upon fiber type and industrial process is compelled not only by the logic suggested by the RCA, but also by the statistics shown

38

in tables 3-3 and 3-4. The "range of uncertainty" around the OSHA coefficients suggests that the coefficients are significantly different from each other. The upper bounds for the McDonald and the New-house and Berry coefficients are 0.0011 and 0.008, far less than the coefficients for the Dement, Finkelstein, and Selikoff studies. In fact, the upper bounds for the McDonald and the Newhouse and Berry coefficients are well below the lower bound for the Dement and Finkelstein coefficients. Thus OSHA's own analysis suggests that the studies should be treated as measuring different risks.

With respect to the health effects of asbestos in buildings, the RCA and the EPA have taken the position that the risks can best be estimated using studies of workers exposed to asbestos insulation in buildings. The only such study is the Selikoff study. Coincidentally, the OSHA estimate of average coefficients yields numbers very close to the Selikoff coefficients estimated by RCA and CHAP.

The conclusion that amosite and crocidolite are more dangerous than chrysotile might logically be extended to exposures in buildings. Unfortunately, the Selikoff study, the only study of insulation work-ers, does not include information on the *type* of asbestos to which the workers were exposed. It is likely that most of the cohort worked with a variety of insulation products, and were thus exposed to all types of asbestos. Without data on separate cohorts exposed to the three types of asbestos in building insulation, it is not possible to estimate separate dose-response coefficients for them. Perhaps the relative degree of hazard is in proportion to the relative degrees of hazard for the three types of asbestos in manufacturing, as shown in tables 3-3 and 3-4. We cannot, however, be sure that this rela-tionship is maintained in building situations. I am inclined to believe that the risks caused by exposures in buildings do vary by type of asbestos, so that using a single model will overestimate the risks associated with exposure to chrysotile and underestimate the risk associated with exposure to amosite and crocidolite. However in the absence of more specific data, the modeling of health risks in build-ings must be based on the single set of coefficients derived from the Selikoff study.

Simulated Disease Risks

Risks to Building Occupants

Each study contains simulations of the health effects of low level exposures to asbestos either as a direct calculation of the risk to building occupants, or in a calculation of the single risk faced by an individual exposed to any low level exposure. The RCA uses the Selikoff coefficients from tables 3-3 and 3-4. CHAP uses a range of coefficients differing by a factor of 10, shown in the bottom of tables 3-3 and 3-4. OSHA and NAS use the average coefficients shown in the bottom of tables 3-3 and 3-4. The EPA uses the Selikoff data, with coefficients of $K_L = 0.01$ and $K_M = 15 \times 10^{-9}$. Doll and Peto use their textile manufacturing coefficients.

The results of the simulations of building health risks are shown in table 3-5, with mortality expressed as the lifetime mortality per million exposed persons. In general a mixed cohort of smokers and nonsmokers, males and females is assumed, with the exception that the OSHA and Doll and Peto cohorts are all male. The RCA assumes ten years of exposure to 0.001 f/cc starting at age twenty-two, and finds a risk of sixteen deaths per million. CHAP assumes ten years of exposure to 0.01 f/cc starting at age twenty, a ten-fold increase in the dose, and finds a risk of 15 to 150 deaths per million. The CHAP estimate may be divided by ten to determine the predicted mortality for an exposure intensity of 0.001 f/cc. This yields mortality of 1.5 to 15, somewhat less than that of the RCA. The CHAP model would be expected to produce lower estimates than the RCA model prin-

Table 3-5. Predicted Lifetime Asbestos Mortality Risks for Building Occupants: Various Studies

Report	Exposure f/cc	Duration (years)	Age at first exposure	Cohort[a]	Mortality per million exposed persons
RCA model	0.001	10	22	M,F;S,NS	16
	0.001	10	7	M,F;S,NS	31
CHAP (1983)[b]	0.01	10	20	M,F;S,NS	15–150
OSHA (1983a)[c]	0.1	20	25	M;S,NS	2,260
Putnam, Hayes, and Bartlett (1984)	0.008	35	25	M,F;S,NS	140[d]
NAS (1984)[e]	0.0004	73	0	M,F;S,NS	33
Doll and Peto (1985)[f]	0.0005	20	20	M;S,NS	10

[a] M = male; F = female; S = smokers; NS = nonsmokers.
[b] p. 131.
[c] p. 52c.
[d] The Putnam, Hayes and Bartlett study defines an exposed population, but over the 35 year period, individuals are retired from the cohort, and new individuals enter. We have assumed that the total number of exposed persons is twice the population exposed at any one time.
[e] p. 212.
[f] p. 47.

cipally because less mesothelioma would be predicted by a model with a smaller exponent on time, and with a ten-year latency for mesothelioma.

OSHA's minimum exposure is 0.1 f/cc, yielding mortality of 2,260 for a twenty-year exposure. Dividing this by 200 to make the exposure comparable to that of the RCA would yield 11.3 deaths, similar to the RCA estimate. This follows from using a mesothelioma model that incorporates latency, and from using a smaller mesothelioma coefficient than RCA. The Putnam, Hayes, and Bartlett estimate for the EPA is difficult to compare because it does not appear to be based on a single cohort. The NAS assumes a lifetime exposure to 0.0004 f/cc, and finds mortality of thirty-three. This estimate cannot be easily adjusted for comparability with the RCA estimate because of the lifetime exposure assumed by NAS. The coefficients used by NAS are almost twice those used by the RCA, so it seems likely that the NAS model would predict considerably more disease than the RCA model at comparable exposures. On the other hand, the NAS mesothelioma model employs an exponent in equation (3) of only 3.2, so it will tend to estimate less mesothelioma for early first exposures.

This review of the results of the simulations indicates that they do not differ greatly. In fact, the results may differ less than the coefficients. This would be expected. Because the models differ in some details, estimating the coefficients separately for each model yields different coefficients. In any event, it appears that the greatest disease risks are predicted by the RCA model and, for exposures that do not begin at a young age, by the NAS model. Thus although the RCA model differs crucially from some of the others as to whether risks differ by fiber type and process, this disagreement does not have a great impact on the estimate of risks to building occupants.

Table 3-6 explores the effect of various conditions on mortality risks in buildings, using the RCA model. The top half of the table shows predicted lives and life-years lost of a cohort of one million persons, half of whom are smokers (or in the case of seven-year-olds will become smokers). Increasing the duration of exposure increases health risks less than proportionally, since early years of exposure cause a greater mesothelioma risk than do subsequent years of exposure. Decreasing the age of first exposure increases risks. A cohort first exposed at age seven experiences risks about double those of a cohort first exposed at age twenty-two. Similarly, increasing the age of first exposure to thirty-five decreases the risk by about 40 percent. The bottom of the table presents mortality risks for a cohort consisting entirely of nonsmokers. The probable number of life-years

Table 3-6. Predicted Lifetime Asbestos Mortality Risks for Building Occupants: RCA Model, Various Conditions

	Duration (years)	Age at first exposure		
		7	22	35
50% smokers, 50% non-smokers		(———lives lost———)		
	1	3.78	1.89	1.03
	10	30.9	15.6	9.04
	20	50.3	26.1	16.0
		(———life-years lost———)		
	1	64.3	27.4	13.2
	10	503	218	114
	20	787	354	197
100% nonsmokers		(———life-years lost———)		
	1	71.7	24.9	8.2
	10	514	183	60.4
	20	775	269	91.4

Note: Assumptions—Exposure intensity = 0.001 *f/cc*; one million persons exposed; cohort is 50% male, 50% female.

lost is reduced by 16 to 25 percent for persons first exposed at age twenty-two, not at all for those exposed at age seven, and is reduced by about 50 percent for those first exposed at age thirty-five. Not smoking fails to benefit those first exposed at age seven because their primary risk arises from mesothelioma, not lung cancer, and since nonsmokers live longer than smokers, they contract *more* mesothelioma. Thus both age at first exposure and smoking habit can significantly affect health risks to building occupants.

Workplace Exposure

The models described above may also be used to predict the risk of mortality faced by workers in various workplaces. Table 3-7 summarizes some projections by the RCA and by OSHA of the risks faced by 1,000 workers. The RCA projections are all based on exposures lasting ten or twenty-five years, while the OSHA projections consider exposures lasting one, twenty, and forty-five years. An exposure intensity of 2 f/cc is assumed throughout. The age at first exposure is twenty-two years for the RCA and twenty-five years for OSHA. The RCA assumes a mixed workforce that is 23 percent female, while the OSHA workforce is all male.

Table 3-7 shows that the workplace risks predicted by the RCA vary by more than two orders of magnitude depending upon the type of exposure, while OSHA assumes the same risk relationship for all types of exposure. The lung cancer risk increases approximately in proportion to the duration of exposure, while the mesothelioma risk, reflecting the nonlinear mesothelioma model, increases less than in proportion to the duration of exposure. Thus the OSHA model predicts that a one-year exposure will yield almost equal amounts of mesothelioma and lung cancer, while a 45-year exposure causes one-third as much mesothelioma as lung cancer. The RCA model predicts no mesothelioma for some types of exposure, such as chrysotile mining and chrysotile manufacturing. Other results from both sets of models, not shown here, reveal that mortality risks have a linear relationship to exposure intensity at low risks, but that as mortality risks rise above 1 percent, risks begin to rise more slowly than exposure.

Table 3-7. Predicted Lifetime Asbestos-Related Cancer Mortality for Asbestos Workers

Exposure	Study	Duration (years)	Mortality risks 1,000 workers exposed to 2.0 f/cc			Life-years lost
			Lung cancer	Mesothe- lioma	Total	
RCA Projections[a]						
Mining	McDonald	10	0.7	—	0.7	9.0
		25	1.8	—	1.8	22.3
Manufacturing, chrysotile						
General	Berry	10	0.9	—	0.9	11.3
		25	2.3	—	2.3	27.9
Textiles	Dement	10	62	—	62	782
		25	144	—	144	1845
Manufacturing, crocidolite	Finkelstein	10	56.8	238	295	4193
		25	127	332	460	6411
Insulation	Selikoff	10	15.5	17.0	32.5	431
		25	37.9	25.8	63.7	819
OSHA Projections[b]						
		1	1.44	1.38	2.96[c]	—
Average of studies		20	27.13	14.08	43.92[c]	—
		45	44.16	15.54	64.11[c]	—

Sources: RCA (1984, pp. 496–499); OSHA (1983, p. 52c).
[a] Age of first exposure is 22 years. The cohort is 77% male, 23% female.
[b] Age at first exposure is 25 years. The cohort is all male.
[c] OSHA totals include gastrointestinal cancer calculated as 10% of lung cancer.

Choosing between the RCA model and one of the other models is of modest significance for predicting building risks, as was shown in the last section. This choice has great significance, however, for predicting risks from various workplace situations, as table 3-7 shows, since the RCA dose-response coefficients vary considerably by fiber type and industrial process, while OSHA uses the same coefficients for all exposure situations.

Comparison with Other Risks

Determining whether risks of a particular disease are large or small is a matter of judgment. It is not clear what other risks ought to form the basis of comparison with asbestos diseases. Rather than try to resolve that issue here, we present, in table 3-8, data on mortality risks faced in a variety of situations. Risks of accidental death are presented in the top of the table, followed by the risk of intentional death by suicide and homicide, and finally by disease risks. The risk caused by exposure in a building for ten years at 0.001 f/cc is presented for comparison. All figures in the table are expressed as the risk of death in any year per 100,000 population. The asbestos risk is arrived at by converting the lifetime risk per million exposed persons (15.6) to an annual risk per 100,000 persons. This conversion is achieved by assuming an average life span of seventy years, and dividing the lifetime risk per million persons by 700.

The risk of cancer mortality from exposure to 0.001 f/cc in a building is about one one-thousandth the risk of suffering an automobile-related fatality, and one five-hundredth the risk of being a victim of suicide or homicide. It is one one-hundredth the risk of drowning, and one-fortieth the risk of dying in a firearm accident. Turning to disease risks, the risk faced by building occupants is one one-thousandth the risk of dying of pneumonia, and less than one-thirtieth the risk of dying from tuberculosis. Thus even for those who are exposed to asbestos in buildings for a decade, the risk of ultimately dying from asbestos-related cancer is well below the risk of dying from common accidental causes or of diseases that are thought to be uncommon in North America today.

Another way to view these risks is to compare two risks associated with building occupancy itself. If a person works for ten years in a building in which he is exposed to an airborne asbestos concentration of 0.001 f/cc, his lifetime risk of premature mortality is 15.6×10^{-6}. Suppose that this person drives 5 miles each way to the building

Table 3-8. Risk of Death by Cause, United States, 1983 (mortality rate per 100,000 population/year)

Accidental deaths[a]	
Motor vehicle[b]	19.1
Falls	5.0
Fires, burns	2.0
Drowning	2.8
Firearms	0.8
Choking	1.4
Poison gas	0.6
Intentional deaths[c]	
Suicide	12.3
Homicide	10.7
Disease[c]	
Heart disease	325.8
All cancers	188.8
Pneumonia	22.3
Diabetes mellitus	15.1
Emphysema	5.7
Chronic bronchitis	1.5
Asthma	1.4
Tuberculosis	0.8
Influenza	0.6
Asbestos disease from building exposure	
RCA model (0.001 f/cc: 10 yrs)	0.022

[a] National Safety Council (*World Almanac*, 1985, p. 700).
[b] Between 1980 and 1982, passengers in autos and taxis averaged 1.2 fatalities per 10^8 passenger-miles. National Safety Council.
[c] National Center for Health Statistics, U.S. Department of Health and Human Services.

250 days per year. The fatality rate for automobile and taxicab occupants in the United States in 1980–1982 was 1.2 per 10^8 passenger miles. The risk that this commuter would die in an auto accident travelling to and from the building is 300×10^{-6}. Thus the risk of dying decades later from asbestos-related disease is one-twentieth the risk of dying in an auto accident while commuting to and from the building during the ten-year exposure period. Since driving causes risks for pedestrians and others as well as for motor vehicle occupants, the total risk of an accidental fatality is greater than the risk to the vehicle occupants alone. The RCA concluded that the total risk of a fatality from commuting ten miles each day was fifty times the risk of dying from an asbestos-related disease (RCA, 1984, p. 585).

Another set of comparisons may be made with risks of cancer from other causes. Table 3-8 shows that the risks caused by ten years of exposure to 0.001 f/cc of airborne asbestos in a building are less than one one-thousandth the risk of dying from all forms of cancer. The current occupational exposure limit for asbestos exposure in the United States is 2.0 f/cc.[1] Suppose that actual worker exposures average half the exposure limit.[2] The OSHA model predicts that workers exposed to 1.0 f/cc for ten years would experience a lifetime risk of asbestos-related mortality equal to about 11,000 per million persons exposed (OSHA, 1983a, p. 52c). This is almost 1,000 times the risk predicted by the same model for the building exposure. The current occupational exposure limit for chrysotile asbestos exposure in Ontario is 1.0 f/cc. The RCA model predicts that workers exposed under this limit for ten years would experience a lifetime risk of mortality equal to 200 per million (RCA, 1984, p. 428). This is more than twelve times the risk presented by the building exposure. The occupational exposure limit for benzene is five parts per million in Ontario, implying risks for those exposed over their working lifetimes of 25,000 fatalities per million exposed.[3] The risks faced by building occupants thus seem small by comparison to the formal regulations governing workplace exposures to carcinogens.

[1] In June 1986, OSHA published a final rule reducing this exposure limit to 0.2 f/cc. 51 *Federal Register*, 22612, June 20, 1986.

[2] If an exposure limit is only to be exceeded rarely, the *average* exposure will be one-half that limit, or less. See RCA (1984), p. 382.

[3] Extrapolation from McGrath and Kusiak (1983), app. C. In the United States, OSHA reduced the benzene limit from 10 to 1.0 parts per million in 1977, but this lower limit was invalidated by a court challenge. See Lave (1981) p. 99. A reduced limit has been proposed again. 50 *Federal Register*, 50512, Dec. 10, 1985.

REFERENCES

Berry, Geoffrey and Muriel L. Newhouse. 1983. "Mortality of Workers Manufacturing Friction Materials Using Asbestos," *British Journal of Industrial Medicine* vol. 40, no. 1 (February) pp. 1–7.

CHAP. *See* Chronic Hazard Advisory Panel on Asbestos.

Chronic Hazard Advisory Panel on Asbestos. 1983. *Report to the U.S. Consumer Product Safety Commission* (Washington, D.C., U.S. Consumer Product Safety Commission, July).

Dement, John M., Robert L. Harris Jr., Michael J. Symons, and Carl M. Shy. 1982. "Estimates of Dose-Response for Respiratory Cancer Among Chrysotile Asbestos Workers," *Annals of Occupational Hygiene* vol. 26, no. 1–4, pp. 869–887.

Doll, Richard, and Julian Peto. 1985. *Effects on Health of Exposure to Asbestos* (London, Her Majesty's Stationary Office, U.K. Health and Safety Commission).

Finkelstein, Murray M. 1983. "Mortality Among Employees of an Ontario Asbestos-Cement Factory" (Toronto, Ontario Ministry of Labor, February; revised September).

Hammond, E. Cuyler, Irving J. Selikoff, and Herbert Seidman. 1979. "Asbestos Exposure, Cigarette Smoking and Death Rates," *Annals of the New York Academy of Sciences* vol. 330 (December 14) pp. 473–490.

Henderson, Vivian L. and Philip E. Enterline. 1979. "Asbestos Exposure: Factors Associated with Excess Cancer and Respiratory Disease Mortality," *Annals of the New York Academy of Sciences* vol. 330 (December 14) pp. 117–126.

Lave, Lester B. 1981. *The Strategy of Social Regulation* (Washington, D.C., Brookings Institution).

McDonald, J. Corbett, F. Douglas K. Liddell, Graham W. Gibbs, Gail E. Eyssen, and Alison D. McDonald. 1980. "Dust Exposure and Mortality in Chrysotile Mining, 1910–75," *British Journal of Industrial Medicine* vol. 37, pp. 11–24.

McGrath, T. and R. Kusiak. 1983. *An Analysis of the Health Effects of Benzene and Recommendations for an Occupational Standard* (Health Studies Services, Ontario Ministry of Labour, July).

NAS. *See* National Academy of Sciences.

National Academy of Sciences, Committee on Nonoccupational Health Risks of Asbestiform Fibers, National Research Council. 1984. *Asbestiform Fibers; Nonoccupational Health Risks* (Washington, D.C., National Academy Press).

Nicholson, William J., George Perkel, Irving J. Selikoff, and Herbert Seidman. 1981. "Cancer from Occupational Asbestos Exposure: Projections 1980–2000," in J. Peto and M. Schneiderman, eds., *Banbury Report 9: Quantification of Occupational Cancer* (Cold Spring Harbor, N.Y., Cold Spring Harbor Laboratory) pp. 87–111.

Nicholson, William J., Irving J. Selikoff, Herbert Siedman, Ruth Lillis, and Paul Formby. 1979. "Long-Term Mortality Experience of Chrysotile Miners and Millers in Thetford Mines, Quebec," *Annals of the New York Academy of Sciences* vol. 330 (December 14) pp. 11–21.

Ontario Royal Commission on Asbestos. 1984. *Report of the Royal Commission on Matters of Health and Safety Arising from the Use of Asbestos in Ontario* (Toronto, Queen's Printer).

OSHA. *See* U.S. Occupational Safety and Health Administration.

Peto, Julian. 1983. "Dose and Time Relationships for Lung Cancer and Mesothelioma in Relation to Smoking and Asbestos Exposure," in *Zur Beurteilung der Krebsgefahren durch Asbest* (Proceedings of the Bundesgesundheitsamt Asbestos Symposium) (Berlin, February 1982. bga Schriften, MMV Medizin Verlag Munchen).

———. 1980. "Lung Cancer Mortality in Relation to Measured Dust Levels in an Asbestos Textile Factory," in J. C. Wagner, ed., *Biological Effects of Mineral Fibres*, vol. 2, IARC Scientific Publications, no. 30 (Lyon, France, International Agency for Research on Cancer) pp. 829–835.

Peto, Julian, Brian E. Henderson, and Malcolm C. Pike. 1981. "Trends in Mesothelioma Incidence in the United States and the Forecast Epidemic Due to Asbestos Exposure During World War II," in R. Peto and M. Schneiderman, eds., *Banbury Report 9: Quantification of Occupational Cancer* (Cold Spring Harbor, N.Y., Cold Spring Harbor Laboratory) pp. 51–72.

Putnam, Hayes and Bartlett, Inc. 1984. "Cost and Effectiveness of Abatement and Asbestos in Schools." Draft (U.S. Environmental Protection Agency, Office of Pesticides and Toxic Substances, August 8).

RCA. *See* Ontario Royal Commission on Asbestos.

Rubino, G. F., G. Piolatto, M. L. Newhouse, G. Scansetti, G. A. Oresini, and R. Murray. 1979. "Mortality of Chrysotile Asbestos Workers at the Balangero Mine, Northern Italy," *British Journal of Industrial Medicine* vol. 36, pp. 187–194.

Seidman, Herbert, Irving J. Selikoff, and E. Cuyler Hammond. 1979. "Short-Term Asbestos Work Exposure and Long-Term Observation," *Annals of the New York Academy of Sciences* vol. 330 (December 14) pp. 61–89.

Selikoff, Irving J., E. Cuyler Hammond, and Herbert Seidman. 1979. "Mortality Experience of Insulation Workers in the United States and Canada, 1943–1976," *Annuals of the New York Academy of Sciences* vol. 330 (December 14) pp. 91–116.

UKAC. *See* U.K. Advisory Committee on Asbestos.

U.K. Advisory Committee on Asbestos. 1979. *Asbestos—Final Report of the Advisory Committee,* 2 vols. (London, Her Majesty's Stationery Office).

U.S. Occupational Safety and Health Administration. 1983a. *Quantitative Risk Assessment for Asbestos Related Cancers* (Mimeo) (Washington, D.C., OSHA, October).

_____.1983b "Occupational Exposure to Asbestos; Emergency Temporary Standard. 29 CFR Part 1910," *U.S. Federal Register* vol. 48, no. 215, 51086–51140, November 4.

Viscusi, W. Kip. 1982. "Setting Efficient Standards for Occupational Hazards," *Journal of Occupational Medicine* vol. 24, no. 10 (December) pp. 969–976.

Walker, Alexander M., J. E. Loughlin, E. R. Friedlander, K. J. Rothman, and N. A. Dreyer. 1983. "Projections of Asbestos-Related Disease 1980–2009," *Journal of Occupational Medicine* vol. 25, no. 5 (May) pp. 409–425.

Weill, H. 1979. "Influence of Dose and Fiber Type on Respiratory Malignancy in Asbestos Cement Manufacturing," *American Review of Respiratory Disease* vol. 120, pp. 345–354.

four

CONTROL OF ASBESTOS IN BUILDINGS

Methods of Control

Once asbestos-containing friable material (ACFM) has been discovered in a building, any of four control actions may be taken: maintenance and custodial control, encapsulation, enclosure, or removal. Each of these control measures is described briefly in this section.[1] The next section outlines current regulations and practices regarding control actions. The last two sections summarize the current state of knowledge regarding the costs and effects of these control actions and describe the effects of exposure control.

Maintenance and custodial control, also referred to as "special operations and maintenance procedures and periodic reassessment" and "deferred action," is appropriately instituted any time that ACFM is found in a building. It involves recording the location of any ACFM in the building and carefully cleaning up any asbestos that may have been released from this material. Regular building workers, including maintenance and custodial personnel, are instructed in the location of the material and in safe methods of working in the vicinity of it. The material is reexamined periodically to determine whether conditions warrant further corrective action.

Encapsulation involves spraying the ACFM with a sealant that binds the fibers together to reduce the probability of their release. Encapsulation is not a permanent solution in that the ACFM must still be removed from the building at some time. The process of applying the encapsulant can cause high dust levels. In some cases, after encapsu-

---

[1]For more detail, see RCA (1984), pp. 609–616; EPA (1985), chaps. 3–6.

lation patches of encapsulated material may fall loose releasing asbestos fibers from the underlying material. While building owners often prefer it because of its simplicity, encapsulation is recommended only where the ACFM is in good condition and sufficiently thin that it can be fully penetrated by the sealant (RCA, 1984, p. 612).

Enclosure involves covering the ACFM with a structure that will protect it from damage and prevent the release of asbestos fibers. The enclosure may be made of plywood or drywall, or it may consist of cloth or sheet metal for wrapping pipe and boiler insulation. Like encapsulation, enclosure is a temporary measure. The RCA recommended enclosure for ACFM that is difficult to remove, or for pipe and boiler insulation that is in good condition.

Removal with safe disposal is the only permanent solution to an asbestos problem, since this is the only solution in which the ACFM is removed from the building. Removal must be performed carefully to ensure that removal workers are not exposed to excessive airborne asbestos concentrations, that other portions of the building are not contaminated, and that the asbestos is fully removed so that no risk to building workers remains. The removed ACFM must be packaged in leak-proof containers and transported carefully to an appropriate dump site, where care must be taken to bury it without breaking open the containers. After the asbestos material is removed, a non-asbestos insulating material may have to be installed, if the insulating function of the ACFM is still desired.

Current Regulations and Practices

The only explicit government requirement for building owner inspection for ACFM presence is contained in the EPA's "Asbestos-in-Schools" rule, issued in 1982, which requires schools to inspect for friable asbestos and to notify employees and parents if such asbestos is found.[2] Recommendations for complying with this requirement are contained in a report, *Guidance for Controlling Asbestos-Containing Materials in Buildings* (EPA, 1985). The rule does not require that any control action be taken, but the practical effect of notifying employees and parents that a school contains asbestos is to create considerable political pressure to undertake some control action at

[2]"Friable Asbestos-Containing Materials in Schools, Identification and Notification Rule," 47 *Federal Register* 23360, 40 *Code of Federal Regulations*, Part 763.

once. The Asbestos School Hazard Abatement Act of 1984[3] provides loan and grant funding to assist schools in controlling asbestos hazards, replacing asbestos-containing material, and restoring buildings to their pre-control conditions.

Regulations controlling exposures to asbestos in buildings have been promulgated by both EPA and OSHA. EPA's regulations, issued under the National Emission Standards for Hazardous Air Pollutants (NESHAPS) of the Clean Air Act, ban the application in buildings of sprayed, molded, or wet-applied asbestos-containing materials that are friable after installation.[4] The same regulations specify procedures that must be followed to control asbestos fiber release during the renovation or demolition of a building containing friable asbestos.[5] While these regulations will protect renovation and demolition workers, their ostensible purpose is to control the release of asbestos fibers to the outdoor air. EPA has issued several guidance documents specifying appropriate procedures for inspecting a building for asbestos and for controlling that asbestos, but these documents are advisory only, rather than binding. Even they do not state clearly *when* control is needed and when the asbestos may safely be left alone.

OSHA's regulations apply to all workers within OSHA's jurisdiction, but were designed specifically for factory workers. They specify airborne asbestos fiber exposure limits and procedures to be used to achieve these limits.[6] The exposure limit adopted in 1976 was 2.0 f/cc as an 8-hour time weighted average. On November 4, 1983, an Emergency Temporary Standard of 0.5 f/cc was promulgated, but has subsequently been stayed by court order.[7] The application of the OSHA regulation to buildings is controversial on two grounds. First, it is argued that passive building occupants deserve greater protection than is offered to industrial workers who, unlike building occupants, presumably know about the hazards of asbestos and its presence in the work environment when they accept their employment. Second, the optical microscope method, used to analyze airborne asbestos fiber levels in a factory and required by OSHA as the

---

[3]P.L. 98-377, Aug. 11, 1984, 98 Stat. 1287, 20 USC 4011.

[4]40 *Code of Federal Regulations* 61.148 and 61.150, July 1, 1984.

[5]40 *Code of Federal Regulations* 61.147, July 1, 1984.

[6]29 *Code of Federal Regulations* 1910.1001(a).

[7]29 *Code of Federal Regulations* 1910.1001(k). In 1986 OSHA published a final rule incorporating a limit of 0.2 f/cc. 51 *Federal Register*, 22612, June 20, 1986.

only means for testing compliance with its regulations, does not accurately measure asbestos fiber levels in a building where many other types of dust may be present because it cannot distinguish between asbestos fibers and some other types of fibers. During 1984 and 1985, the EPA was subject to vigorous criticism for failing to provide positive guidance as to when control actions are needed, and when it is appropriate to leave the asbestos undisturbed. The EPA guidance documents identify the factors that should enter into this decision, but there is no clear statement of what combination of factors demands action. In part, EPA critics are searching for certainty where it cannot be found. The infinite variety of situations in which ACFM may be found makes it virtually impossible to provide clear guidelines of universal applicability. The other element in EPA's dilemma is that the health hazard from exposure to airborne asbestos fibers is thought to be proportional to exposure, so that there is no "absolutely safe" exposure. To adopt a particular guideline for controlling asbestos in buildings would be to accept implicitly some risk of fatal disease. This EPA is loath to do.

EPA's attitude toward the health risk posed by asbestos in buildings and the urgency of taking corrective action has evolved considerably between the 1983 and 1985 guidance documents. In 1983, 50 percent of airborne asbestos concentrations in schools were said to fall between about 80 and 800 ng/m³ (EPA, 1983, p. 1-5). The text on health effects states in part:

Even though exposure levels are likely to be low in comparison with industrial levels, any additional exposure above background (outdoor) levels should be avoided if possible. A prudent response by building owners requires recognition of the potential hazards and serious consideration of appropriate abatement actions. (EPA, 1983, p. 1-4)

The 1985 Guidance Document states that 50 percent of airborne asbestos concentrations in schools fall between about 1 and 100 ng/m³, one to two orders of magnitude below the 1983 estimates (EPA, 1985, p. 1-4). The text on health effects states in part:

. . . only a small proportion of people exposed to low levels of asbestos will develop asbestos-related diseases. . . . Also, asbestos exposure in children is of special concern: since they have a greater remaining lifespan than adults, their lifetime risk of developing mesothelioma is greater. Avoiding unnecessary exposure to asbestos is prudent. (EPA, 1985, p. 1-2)

As the estimated exposure level has fallen, so, quite sensibly, has the recommended urgency of corrective action. A prudent building owner now should avoid unnecessary exposure, but need not necessarily take abatement action.

In Ontario, spraying of asbestos-containing insulation stopped in 1973, in part because of costly regulations governing the spray process.[8] The RCA recommended that the spraying of asbestos-containing friable material and the application of asbestos-containing friable pipe and boiler insulation be prohibited (RCA, 1984, pp. 553, 557). The RCA also recommended the adoption of a comprehensive program regulating asbestos in buildings (RCA, 1984, chap. 10). There are four main regulatory provisions in this program. First, prior to demolition, renovation, maintenance, or custodial work that will disturb friable material in a building, that material must be tested to determine whether it contains asbestos. Second, if there is asbestos-containing friable material in a building, then a program of maintenance and custodial control must be instituted and followed so long as the material remains in the building. Third, if asbestos-containing friable material is present in a building, then when demolition, renovation, maintenance, or custodial work would disturb that material, specified safe work practices must be followed. Finally, prior to demolition of a building, any asbestos-containing friable material must be removed. Some elements of these recommendations are included in *Regulation Respecting Asbestos on Construction Projects and in Buildings and Repair Operations.*[9]

Costs of Control

Controlling ACFM in a building imposes two types of costs: the cost of working on the asbestos itself, and the cost of dislocating the occupants during this work. The latter set of costs has been generally ignored in the past, probably because most control work to date has involved school buildings. The work is often performed during vacation periods when the buildings would normally be unused or underused, so there is no relocation of occupants. Since moving of furniture is often performed by permanent staff, no special appropriations are needed for this work. When the building is a commercial office building, factory, or warehouse, it may no longer be appropriate to ignore these costs.

[8]Regulation 419/73, sec. 33 under the Ontario *Construction Safety Act*.

[9]Ontario Regulation 654/85, effective March 16, 1986.

54

The costs of controlling the asbestos itself have been the subject of a number of surveys by the EPA and its consultants. These costs generally exhibit a wide range of values, depending upon the condition of the building, the location and type of asbestos material, local labor costs, and other factors. One set of costs based upon a survey of contractors experienced in asbestos abatement is presented in table 4-1. The cost of removal and safe disposal of the asbestos material and replacement with a substitute material ranges from $4.25 per square foot of insulation to $9.14 per square foot. Removal and disposal alone, without replacement, cost $3.31 to $7.71 per square foot. Enclosure and encapsulation are significantly less expensive than removal. It must be remembered, however, that with enclosure or encapsulation there will be an additional cost at some time in the future when the material must be removed from the building. It seems likely that removal at the time of demolition would cost less than removal before reoccupancy of the building because precautions to protect the building from water damage need not be taken. We have no data on the cost of removal prior to demolition to quantify this saving.

Dislocating building occupants may involve costs for moving them out of the building and back, for renting alternate space, and for lost productivity during the move. A survey of moving companies in the Washington, D.C., area revealed a range of moving costs even for a move within a one mile radius. Most companies assumed that each employee (or "position") would occupy 200 square feet of floor space, and the cost estimates ranged from $0.25 to $0.50 per square foot of floor space. Doubling these costs for a two-way move yields the

Table 4-1. Costs Associated with Asbestos Control
(1985 U.S. dollars)

Cost category	Low average	High average	Unit of measurement
Removal, disposal, replacement[a]	4.25	9.14	sq. ft. material
Removal, disposal[a]	3.31	7.71	sq. ft. material
Enclosure[a]	3.50	3.75	sq. ft. material
Encapsulation[a]	1.85	3.51	sq. ft. material
Moving[b]	0.50	1.00	sq. ft. space
Alternate space[c]	1.67	3.33	sq. ft. space
Lost productivity[d]	wage × 0.02		employee

[a] Putnam, Hayes & Bartlett, (1984) pp. 1-6, 1-9.
[b] Two-way move at $0.25 to $0.50 per square foot each way.
[c] Two months at $10 to $20 per square foot per year.
[d] Assumes one week lost work.

figures in table 4-1. The cost of alternate space is based on the assumption that the space will be needed for two months. Office space in the Washington area rents for $10 to $20 per square foot per year, or $1.67 to $3.33 per square foot for the two months. This cost could vary considerably depending upon the availability of suitable space in the vicinity of the building being worked upon. The cost of lost productivity is based on the survey of movers who said that a typical office move occurs over the weekend and that Friday and Monday are generally lost. Adding a half day for sorting out the new quarters and multiplying by two yields one week of lost time. This is valued at one-fiftieth of the annual wage of the workers, here assumed to be $25,000.[10] Once again, the actual cost could vary considerably depending upon how much time is required for the occupants to become fully productive after the move, and upon their average wage.

If control actions are deferred, a program of maintenance and custodial control should be instituted to monitor the condition of the asbestos and ensure that significant quantities of fibers are not released. The nature and cost of this program will depend upon the particular situation, and could vary enormously from one building to another. If the asbestos insulation is in good condition and not being disturbed, the control program may require only that the maintenance and custodial staff be informed of the location of the material, be trained in safe practices for dealing with it, and be required to inspect it from time to time. If, on the other hand, the material is often disturbed because of minor maintenance activities, considerable costs may be incurred to protect workers during these activities and to clean up afterward. No reliable studies of the magnitude of these costs have been found. This study will assume that the control program cost is proportional to the area of ACFM in the building. It is further assumed, rather arbitrarily, that twenty days per year of custodial and maintenance time will be required for special cleaning, inspection, and precautions in a building of 100,000 square feet.

The actual cost of a control program will depend upon its timing. Suppose, for example, that removal is the desired control action,

---

[10]The average annual wage in the District of Columbia in 1983 was $23,864, excluding farm, railroad, and armed forces, according to the U.S. Department of Commerce, Bureau of the Census, *Statistical Abstract of the United States 1985*, p. 418, table 698. This includes both blue collar and white collar workers. The white collar (office) rate should be higher, but a national average would be considerably lower. This value was rounded to $25,000 to represent those cases located in the Washington metropolitan area. In fact, the true loss is the worker's marginal product, which should probably equal his wages plus benefits, a higher number.

and that the condition of the building requires removal at once. This would involve the full cost of removal and reinsulation plus the full cost of dislocating the current tenants. Assuming an average office wage of $25,000, and assuming a building in which the area of ACFM equals the floor area of the building, the total cost is $8.92 to $15.97 per square foot. Suppose removal might be deferred until such time as major renovation of the building occurs. At this time the occupants must be moved out anyway, so the only occupant-related cost is the cost of alternate space. We assume that if the building is being renovated, only one additional month is required for asbestos removal. In this case, the cost of control, ignoring the cost of a maintenance and custodial control program until removal, is $5.08 to $10.81 per square foot, a reduction of about one-third. If removal is postponed until demolition of the building, there are no costs for occupant disturbance, and no cost of reinsulation. The resulting cost is $3.31 to $7.71 per square foot, less than half the cost of immediate removal. These calculations do not discount future costs, as would be required in a present value calculation. They show substantial savings from deferring action, but only on the assumption that the cost of caring for the asbestos in the interim is small.

Effects of Control Work

The effects of control work on the exposure of building occupants and workers can be determined only by careful measurement of airborne asbestos fiber levels in buildings both before and after the control work has been performed. The meager data available on airborne asbestos fiber levels has already been lamented in this report. It should not be surprising that before and after studies of airborne asbestos levels are rarer still. Pinchin's study for the Ontario RCA revealed low fiber levels both before and after, and no significant reduction in the average fiber level (Pinchin, 1982, p. 7.12). In fact, the fiber level was often higher after control work than before. Sebastien (1977, p. 64) found increased airborne asbestos levels after removal, but considerable reductions after encapsulation and enclosure. Other studies, using varied measurement methods, present mixed results. It seems likely that high quality control work in a building that initially experiences high airborne dust levels should generally cause a considerable reduction in those dust levels. Conversely, careless work, or work performed in a building with low initial asbestos dust levels, may fail to decrease, or actually increase, airborne asbestos concentrations.

For purposes of this study, we will assume that control work, if well done, will reduce occupant exposures by 100 percent. This has the advantage of simplicity, but it overstates the actual reduction in occupant and worker exposures that would accrue from a real control program.

REFERENCES

Ontario Royal Commission on Asbestos. 1984. *Report of the Royal Commission on Matters of Health and Safety Arising from the Use of Asbestos in Ontario* (Toronto, Queen's Printer).

Pinchin, Donald J. 1982. "Asbestos in Buildings," *Royal Commission on Asbestos Study Series*, no. 8 (Toronto, Ontario Royal Commission on Asbestos).

Putnam, Hayes and Bartlett, Inc. 1984. *Cost and Effectiveness of Abatement of Asbestos in Schools*. Draft for EPA Office of Pesticides and Toxic Substances, August 8.

RCA. *See* Ontario Royal Commission on Asbestos.

Sebastien, Patrick, M. A. Billion-Galland, G. Dufour, and J. Bignon. 1980. *Measurement of Asbestos Pollution Inside Buildings Sprayed with Asbestos.* Translation of a document prepared for the Government of France, Ministry of Health and Ministry for the Quality of Life Environment, 1977, EPA 560/13/80-026 (Washington, D.C., U.S. Environmental Protection Agency, August).

U.S. Environmental Protection Agency, Office of Toxic Substances. 1985. *Guidance for Controlling Asbestos-Containing Materials in Buildings,* 1985 Edition, EPA 560/5-85-024 (Washington, D.C., U.S. Environmental Protection Agency, June).

five

ECONOMIC FRAMEWORK

The problems raised by the presence of hazardous materials in buildings are of several types that may be analyzed using distinct economic models. The first problem is to determine the effect on the timing of building demolition of a requirement that a hazardous material must be removed just before demolition of the building. Here we use an economic model that ignores tax considerations in order to explore the problem with minimum complexity. The second problem is to perform a social evaluation of the cost-effectiveness of various policies concerning the treatment of hazardous materials in buildings. We use a cost-effectiveness model in which future costs and effects are discounted to the present to calculate the cost per life-year saved of the various control policies. The third problem is to determine the costs perceived by a building owner who has discovered a hazardous material in his building and must now remove the material or ensure that it does not present unacceptable hazards to occupants and workers while it remains in the building. We use a financial model of a firm, including taxes and depreciation, to determine the effect that discovery of a hazardous material and required subsequent control policies many have on the value of the firm. Each of these models is described in more detail in the following sections and in appendixes A and B. Readers who are not interested in technical economic matters may wish to skip this chapter.

Material Control Costs and the Timing of Demolition

Suppose that in certain cases, the hazardous material may remain in the building until that building is demolished, but must be removed prior to demolition. The required removal prior to demolition has the effect of increasing the cost of demolition. What effect will this have on the economic life of the building, that is, the time until demolition?

The problem of deciding when to replace a particular piece of capital is explained by Salter (1960). He analyzes investment in more productive capital by analyzing the decisions of a single firm regarding a single piece of equipment. He concludes that a plant should be abandoned when the earned surplus over operating costs fails to yield a normal rate of return on the current value of the site, related working capital, and scrap value (Salter, 1960, p. 56). This criterion recognizes that sunk costs are sunk, and that depreciation is irrelevant to the investment decision. When replacement of the capital is contemplated, the criterion is whether the present value of expected future surpluses over operating cost of the new plant exceeds the expected future surpluses over operating cost of the existing plant by an amount sufficient to earn a normal rate of return on the investment net of the scrap value of the old plant.

Turning now to the model, we assume that revenues and costs associated with the building and the value of the site may grow over time, but that costs grow more rapidly than revenues. To simplify the problem, assume that in general the firm's discount rate exceeds the growth rate of revenue, cost, or site value. Assume further that as the building ages, costs rise at least as rapidly as revenues. Over time, annual profits will fall until at some time T it is more profitable to demolish the building than to retain it. The cost of removing the material just prior to demolition is A. The building owner's discount rate is r. The owner wishes to choose the time of demolition T to maximize the sum of the present value of the profit earned until demolition plus the present value of the building site after demolition.

Appendix A shows that T is an increasing function of A, which means that increasing the cost of material removal postpones the optimal time of demolition of the building under either of two conditions: if revenue and the value of the site are constant while costs rise, or if revenue and costs are constant while the value of the site rises.

This model has ignored income taxes and the sources of corporate financing. A move toward realism would include corporate income taxes in the function to be maximized by the firm, and in the derivation of the firm's discount rate. Because deductions in calculating the corporate income tax effectively subsidize expenses, the after-tax cost of removing the hazardous material is less than the cost if taxes are ignored (see appendix A). Thus the need to remove the hazardous material should retard demolition less in a world with taxes than in a world without taxes.

We have shown that discovery of a hazardous material in a building previously thought to be free of it is likely to cause the optimal date of demolition to be postponed. This is true if the material must be removed at demolition, but only if net revenues are unaffected by the discovery of the hazard. The addition of material removal costs reduces the opportunity cost of continuing to use the building in its present form, and thus allows that use to continue longer along the declining profit stream.

If, however, discovery of the hazard reduces revenue or increases costs, as well as increasing demolition costs, the effect on the timing of demolition is uncertain. Suppose that the revenue schedule shifts down and A increases upon discovery of the material. The effects of these changes will work in opposite directions, and the net effect on T depends upon the magnitudes of the changes. For T to be unaffected, $R + rA$ must be unchanged. If the decline in R exceeds the increase in rA, then the optimal time to demolish will move closer; that is, T will decrease.

Social Cost-Effectiveness Model

The social impact of controlling hazardous material in a building may be modeled by cost-effectiveness analysis. This analysis estimates the cost of achieving a given effect; in the present model, the goal is to reduce the risks of contracting disease caused by a hazardous material in a building. Such analysis raises two questions. First, how should this risk reduction be measured? Should we look at lives saved, life-years saved, or something else? Second, how should we deal with the long time that may elapse between the exposure to the material and the onset of disease? May future disease be discounted, and if so, at what rate? We will consider each of these questions in turn.

Health Effects Measures: Lives Versus Life-Years Lost

How should the health effects of hazardous material in buildings be measured? Several models have been used to evaluate the benefits from reducing the probability of premature mortality. These models assume that the harm from increased risks of mortality can be measured by the willingness-to-pay of the individual plus the willingness-to-pay of others whose utility may be affected by the individual's death. They ignore the willingness-to-pay of others, and the analysis focuses on the affected individual's utility. Cropper (1977) developed a model of decision making regarding occupational health and safety, treating human health as a capital stock, following the work of Grossman (1972). Cropper assumes that workers maximize the discounted value of expected utility subject to a dose-response function for the health hazard, and an income function that increases with greater health hazard exposure. Ippolito (1981) and Freeman (1979, chap. 7) also assume that individuals maximize expected utility EU, with Freeman's model assuming that individuals maximize:

$$\text{Max } EU = \sum_{t=1}^{T} q_t^* \, D^t \, U(x_t) \tag{1}$$

where q_t^* = probability of surviving from period 1 to period t, considering all risks

D^t = $1/(1+r)^t$

r = subjective rate of discount

$U(x_t)$ = instantaneous utility from consuming x_t

The commodity x in equation (1) is a Hicksian aggregate commodity or numeraire, and does not itself present any risk. Thus equation (1) simply presents the individual's willingness to pay for survival so that he may enjoy future consumption. The difficulty in applying this model empirically arises because we have no information about optimal consumption paths x_t and no information about the shape of $U(x_t)$ as an argument. While with perfect foresight and perfect capital markets, marginal utility $U'(x_t)$ must increase annually in proportion to the discount rate,[1] this does not define the ratio of the *total* utilities in the two years.

This leaves the derivation of a simple measure of the value of reducing health risks unresolved. How should these benefits be measured, given our inability to measure utility directly? Much of the literature on the compensating wage differential has calculated a cost per life saved for occupational safety programs, implicitly assuming that the number of lives saved by a policy is an appropriate

[1]In equilibrium with perfect capital markets, an individual can trade present consumption for future consumption at the prevailing interest rate. He will choose to increase current consumption at the expense of next year's consumption until the marginal utility of next year's consumption equals $(1+r)$ times that of current consumption.

index of its reduction in health risk to workers (Thaler and Rosen, 1975; Smith, 1976; Viscusi, 1979). Other studies have discussed the effect of age on the willingness to accept risks. This suggests an additional measure of the loss occasioned by premature death, life-years lost. Is this measure justified?

We can derive these two simple measures of the value of reducing health risks from the model of equation (1) by choosing two simple sets of assumptions about $U(x_t)$. First, suppose that the utility function is expected to be the same in all years of life. This assumption is an extension of the common assumption that preference functions are constant, although it is not a necessary extension since one might have assumed that some periods of life are more enjoyable than others.[2]

Second, assume that consumption x_t is the same in all years of life. This might occur if we assume constant income and no saving. Note that it is *not* a natural outgrowth of maximizing equation (1), since because of time preference this would normally lead to greater consumption early in life and less consumption later in life. These assumptions mean that $U(x_t)$ is the same for all t.[3] Equation (1) may then be rewritten as:

$$\text{Max } EU = U(x) \sum_{t=1}^{T} q_t^* \, D^t \tag{2}$$

Any reduction in the probability of surviving to year t reduces expected utility by an amount proportional to the discounted reduction in the expected years of life lost. Since the value of each life-year is the same, a consistent measure of the health effects of a policy is the number of discounted life-years saved.

Alternatively, assume that the horror of death from disease caused by the hazardous material is far greater than the loss of utility from failing to survive to enjoy future consumption. If the disutility from this form of death is sufficiently great, then we may ignore equation (1) entirely and concentrate on the discounted probability of a death

[2]Stable preferences might mean that an individual can at the present time state his preference function $U(x_t)$ for any future year, and that the realized preferences will coincide with expected preferences. This would not require that $U(x_t)$ be the same for all t. A stronger assumption would be that $U(x_t)$ is identical for all t.

[3]Historically real incomes have increased over time. These increases may or may not be anticipated by individuals at the time that they make decisions about exposure to hazardous substances. Thus the expected time path of consumption may or may not allow for future increases in real income.

itself. In this case, the measure of the health effect of a policy is simply the reduction in the number of asbestos-related deaths discounted to the present from the time of occurrence.

While both of these measures require strong assumptions, we find the former more plausible than the latter. This paper will therefore measure mortality risks in terms of the number of life-years lost.

Discounting Future Effects

The utility model presented in equations (1) and (2) explicitly incorporates a discount rate r for evaluating future consumption. While the inclusion of this factor does not require that it be greater than zero, such values are usually assigned when the discount rate is used. The effect of incorporating a value of r greater than zero is that future mortality risks count less than current risks. Some people object to the discounting of future mortality risks on the grounds that a life is a life, and is worth no less because it is lost in the future than if it is lost now. There are, however, two arguments for applying the concept of discounting to future mortality.

First, it is clear that individuals discount *consumption*. Individuals borrow and lend at positive interest rates in financial markets, and thereby trade future and present consumption at the going interest rate. If future consumption is discounted, it is inconsistent not to discount the loss of future consumption that is associated with premature mortality. Furthermore, if one observes the casual way in which individuals often treat risks that are remote in time, it is hard not to accept that some process like discounting is at work.

Second, discounting future mortality risks in a cost-effectiveness calculation is computationally equivalent to calculating the discounted future value of the costs of avoiding those risks. The future value per unit of effect, computed at the time of the effect, of a stream of costs C_t yielding an effect (for example, reduced mortality risks) E_T in year T is:

$$FV = \left[\sum_{t=1}^{T} C_t \, (1+r)^{(T-t)} \right] / E_T \tag{3}$$

The present value of the same costs and effects, discounting all values to the present is:

$$PV = \left[\sum_{t=1}^{T} C_t / (1+r)^t \right] / [E_T / (1+r)^T] \tag{4}$$

But multiplying the numerator and denominator of equation (4) by $(1+r)^T$ yields equation (3), so they are equivalent.

Consider the following example. Suppose that the expenditure of $1 million in 1985 will reduce the expected loss of life thirty years hence by one life. In 2015 when the life is saved, we can ask what it cost to save that life. The answer is not $1 million, but $1 million plus the interest forgone during the thirty years between 1985 and 2015. If we arbitrarily assume a real interest rate of 4.6 percent, the total cost is $3.85 million. There can be no disputing this result, since spending the money in 1985 rather than 2015 clearly forfits the interest that could otherwise have been earned during that period. But if the future cost of the policy is $3.85 million per life, the present value must be the same. The cost per life saved of the policy in 1985 is $1 million divided by the discounted lives saved, or 0.26 lives (discounted at 4.6 percent), yielding a cost in 1985 of $3.85 million per discounted life saved. In summary, discounting future lives saved back to the present is computationally equivalent to calculating the future value of costs incurred now to save those lives. Since the latter is clearly correct, the former must be also.

The issue of discounting lives is conceptually separate from the question whether individuals will in the future regret decisions that they make today regarding future risks. Freeman (1979, p. 181) notes that individuals may accept today a risk that will come to fruition thirty years hence, only to wish to reject that risk in thirty years for a price larger than the current price plus accumulated interest. Such behavior is inconsistent with a model of fixed preferences, but is not inconceivable, and has been observed in experimental situations. Where inconsistent valuations of life are observed, which one is correct? There is no clear answer to this question. This inconsistency can, however, be dealt with separately from the question of the appropriate discount rate to apply to valuing future health effects.

A further problem arises because in the past income has increased steadily over time. Since the willingness to pay to reduce risks of mortality is probably positively related to income, the benefit models might be modified to reflect increasing income. Suppose that income were expected to increase by a factor of h each year. This might have either of two different effects on lifetime consumption, depending upon the assumptions made. First, suppose that capital markets are perfect, so that individuals can borrow against future income, and that increases in income are fully anticipated. In this case, expected lifetime income affects the level of consumption but not its time profile. Equation (2) may still be used, although consumption levels are in all years greater than if no increase in income was expected.

Second, and in the alternative, assume that either individuals cannot borrow against future income, or that increases in income are not anticipated. In this case, consumption will increase such that $x_t = x_0(1 + h)^t$. Later years of life would be more valuable than early years, but a replacement for equation (2) cannot be specified without knowing the functional form of the relationship between consumption levels and utility.

Under either assumption, willingness to pay to reduce mortality risks measured in the 1970s would understate the willingness to pay that would be expressed decades later when incomes have risen and the health risks of exposure to hazardous materials in the 1980s come to fruition. The studies of the compensating wage differential for exposure to increased accident risks during the 1970s would, if repeated several decades hence, yield higher values. The rate of increase in this value might be approximated by the expected rate of increase in real personal incomes, although this is only an approximation to an unknown function.

Finally, the risk expected from a given exposure to a hazardous substance may change over time. Improved medical knowledge may lead to revisions of the dose-response relationship. Improved medical treatment may reduce the probability of death resulting from a particular disease. In either case, current estimates of the benefits of reducing exposure to the substance would be in error.

The cost of saving life-years by various policies may be compared to several empirical measures of the willingness-to-pay to reduce mortality risks. First, there are data on amounts awarded in tort lawsuits involving accidental death. These amounts are generally based on the loss of future earnings, and are less than the amount an individual would pay to avoid an increase in risk. Second, there are estimates of the compensating wage differential which workers demand to face increased mortality risks. Finally, there are data on the amounts spent in connection with other public programs to protect life, which may be taken to represent the marginal social value of a life-year.

The Choice of a Discount Rate

If one accepts that future health effects should be discounted, what discount rate should be used? The question is not easily answered, as evidenced by a long and contentious literature on the subject.[4]

[4]See Mishan (1975, chap. 33) for a discussion of factors determining the social discount rate. See Fisher and Krutilla (1975, p. 280) for a survey of some of the literature on discount rates.

One might turn to the individual's rate of time preference, to the rate of return to private sector investment, to a social rate of time preference, or to some other rate specifically relevant to future health risks. These rates differ in part because taxes drive a wedge between the pre-tax rate of return to the private corporate sector and the after-tax return realized by the private investor. It may also be argued that the discount rate used by governments for evaluating public policies should be lower than observed private discount rates. The discussion in the preceding section, which argues that discounting health risks is equivalent to calculating the future value of the costs, implies that the social rate of discount is the appropriate benchmark. Lave (1981, p. 44) argues that the social discount rate is not appropriate for valuing future health states because there is not likely to be a fixed transformation between dollars and health.

The present study cannot resolve the venerable debate over the choice of the appropriate discount rate, any more than it will satisfy those who argue that discounting itself is improper. If one polled economists, one would find recommendations for the application of discount rates ranging from near zero to 6 percent or perhaps more. We will assume that individuals can trade money for investments affecting future health states, and therefore refer to the social opportunity cost of investment as the benchmark for this analysis.

There are at least two empirical approaches to determining the social opportunity cost of the investment. One is to use the cost of capital for the private firm (Peskin and Seskin, 1975, p. 23). In this case, the social cost-benefit analysis would yield a result identical to the analysis of the firm except for the effect of the corporate income tax. Another approach is to use the consumer's rate of time preference as the social discount rate, since this reflects the way in which individuals value consumption in different time periods. Lind (1982) argues that unless the project has a zero covariance with returns to the economy as a whole (that is, beta = zero), it should be assumed that the project risk is the same as that of the economy, and thus the market rate of return faced by consumers, after taxes and adjusted for inflation, is the proper discount rate. Lind concludes that this rate is 4.6 percent.

Using social time preference as the discount rate ignores the fact that the funds for the project will displace other investment that would yield the pre-tax private rate of return. This problem is solved by using a shadow price of capital to adjust the capital costs of the project. Lind (1982, p. 86) concludes that the appropriate shadow price multiplier for capital is 3.8, which means that all capital expenditures are multiplied by 3.8. We will follow Lind in using a social discount rate of 4.6 percent, and a shadow price on capital of 3.8.

Model of the Firm

How do firms evaluate the financial consequences of investment decisions? In general, economists assume that firms behave as if they could evaluate complex profit-maximizing problems, even though we know that they do not directly perform the required economic calculations. So here we assume that the firm uses a decision evaluation method based on the capital-asset pricing model (CAPM). While this model was not developed until the 1960s, and is not explicitly used by many firms today, a successful firm may behave as if it used such a model, justifying the use of that model to simulate the firm's behavior.

An alternative would be to consider a simpler model. For example, it is often suggested that firms use simple rules of thumb such as asking whether the project pays for itself within five years. If a simple model with a five-year time horizon is used, the firm will ignore future health effects of many hazardous materials. In the case of asbestos, for example, no significant disease could be expected until more than ten years after occupant or worker exposure to the fibers, so all health benefits from reducing asbestos dust levels would be far outside the decision-making time horizon. The results of the simple model with a short-time horizon are that the firm would take no precautions to reduce risks to building workers and occupants so long as these precautions involved positive costs. This model therefore need not be pursued further.

The model developed here may be used by a firm for deciding whether or not to invest in controlling hazardous materials in a building. See appendix B for the detailed development of this model. Controlling those materials may involve capital investment K_t in year t, and operating cost in subsequent years. The results of this investment will be to reduce possible liability for illness caused by the hazardous materials at some time in the future.

We assume that the objective of the firm is to maximize the value of the firm. This requires that the firm maximize the present discounted value of the expected after-tax cash flow of the firm. The after-tax cash flow consists of revenues less expense, taxes, and capital investment. The firm discounts future cash flows using the weighted after-tax cost of capital as the discount rate. This cost of capital is itself simply the rate of return paid on equity times the fraction of the capital represented by equity plus the rate of return on bonds times the fraction of capital represented by bonds multiplied by one minus the corporate income tax rate, to indicate the after-tax cost of that deductible bond interest. The cost of equity capital for the firm may be determined using the capital asset pricing model (CAPM) and a measure of the riskiness of this firm in relation to the market as a whole.

Workers are assumed to maximize utility given a set of employment opportunities. If labor markets are competitive, the firm will face a perfectly elastic supply of labor at the market wage rate. Where the work environment exposes the worker to health risks, an informed worker will demand a wage premium or compensating wage differential in exchange for accepting the risk associated with the work (Thaler and Rosen, 1975). Barth studied wages paid to asbestos workers and found that a small compensating differential of about four percent of the wage rate may have emerged between the middle 1960s and the middle 1970s when the risks of asbestos exposure were widely publicized. (Selikoff, 1981, p. 576). The model allows for the possibility that workers may demand compensation for working in a building containing the hazardous material.

The financial model of the firm is used to compare the present value of the firm assuming different policies for controlling hazardous material, different liability regimes, and different assumptions about health effects. Since the cost function is separable, the normal operating costs are the same in all evaluations, and therefore drop out of the analysis, leaving the costs of health protection, wage premiums, and health liabilities as the expenses of interest. Investment in the basic building is also the same in all evaluations, as is depreciation on that building, so that only investment to control the hazardous material must be analyzed. In summary, all differences between policies will be a result solely of revenue changes, investments, expenses, and liabilities associated with the hazardous material and its control.

We assume that investment is financed by increasing the financial capital of the firm in an amount just sufficient to cover the investment. In this case, the investment itself would have a negative impact on cash flow, just equaling the payment for the investment. On the firm's balance sheet the change in liability represented by the financial capital is just offset by the change in value of the asset—the modified building.

With respect to the financial aspects of the firm, we assume that the capital structure is invariant with respect to controlling the hazardous material. Since costs associated with controlling a hazardous material are generally a small fraction of total costs, there is no obvious reason why they should affect the capital structure. We also assume that the financial payoff from investing in the control of hazardous material is similar in its risk characteristics to the payoff from other investments by the firm.

REFERENCES

Cropper, M. L. 1977. "Health, Investment in Health, and Occupational Choice," *Journal of Political Economy* vol. 85, no. 6, pp. 1,273–1,294.

Fisher, Anthony C. and John V. Krutilla. 1975. "Valuing Long-Run Ecological Consequences and Irreversibilities," in Henry M. Peskin and Eugene P. Seskin, eds., *Cost-Benefit Analysis and Water Pollution Policy* (Washington, D.C., The Urban Institute).

Freeman, A. Myrick III. 1979. *The Benefits of Environmental Improvement* (Baltimore, Johns Hopkins University Press for Resources for the Future).

Grossman, Michael. 1972. "On the Concept of Health Capital and the Demand for Health," *Journal of Political Economy* vol. 80, no. 1, pp. 223–255.

Ippolito, Pauline. 1981. "Information and the Life Cycle Consumption of Hazardous Goods," *Economic Inquiry* vol. 19, pp. 529–558.

Lave, Lester B. 1981. *The Strategy of Social Regulation* (Washington, D.C., Brookings Institution).

Lind, Robert C. and coauthors. 1982. *Discounting for Time and Risk in Energy Policy* (Washington, D.C., Resources for the Future).

Mishan, E. J. 1975. *Cost-Benefit Analysis* (London, George Allen & Unwin).

Peskin, Henry M. and Eugene P. Seskin, eds. 1975. *Cost-Benefit Analysis and Water Pollution Policy* (Washington, D.C., The Urban Institute).

Salter, W. G. 1960. *Productivity and Technical Change* (Cambridge, Cambridge University Press).

Selikoff, Irving. 1981. *Disability Compensation for Asbestos-Associated Disease in the U.S.* (New York, Environmental Sciences Laboratory, Mt. Sinai School of Medicine).

Thaler, Richard and Sherwin Rosen. 1975. "The Value of Saving a Life: Evidence from the Labor Market," in Nestor Terleckyj, ed., *Household Production and Consumption* (New York, National Bureau of Economic Research) pp. 265–302.

Viscusi, W. Kip. 1979. *Employment Hazards—An Investigation of Market Performance* (Cambridge, Mass., Harvard University Press).

six

ECONOMIC ANALYSIS OF ASBESTOS CONTROL OPTIONS

The Cases

The issues addressed in this study cannot be resolved without determining empirically the relative magnitudes of a number of variables. Yet buildings containing friable asbestos material constitute an infinite variety of sizes, shapes, uses, and design details. Defining a small set of "representative" buildings for analysis appears to be hopeless. Instead, three buildings will form the basis for three case studies. These buildings are of several types, with asbestos insulation in varying quantities and conditions. All are offices or schools; there are no factories, warehouses, or examples of the many other types of buildings that contain asbestos. It is not suggested that these case studies are typical buildings of their type, except for Case 2 which has the characteristics of an average school building. Many major types of buildings are not represented here.

Because the cases are not offered as representative buildings, any conclusions from the case studies must be drawn carefully. Accordingly, most of our conclusions are specific rather than general. A case may show that a result *can* occur, but not how common is its occurrence. The great variability of the building stock requires that any real building be analyzed on its own; conclusions from each case study are not generally applicable. Sensitivity analyses are performed to determine the extent to which the results are affected by changes in key parameters. Some general principles can be presented that will in some cases allow quick assessment of a building by using a small set of facts, without detailed analysis. In a few instances the results are so strong, and so insensitive to reasonable variations in parameters, that we can confidently suggest a general result applicable to most buildings.

Table 6-1 summarizes some characteristics of the case studies. Case 1 represents an office building in which asbestos insulation is assumed to be sprayed on all ceilings, behind a drop ceiling. Average control cost and exposure data are imputed to the building for the economic analysis. Case 2 is a school building, with all characteristics based on averages derived from an EPA survey of asbestos-containing schools. Case 3 is a professional building with sprayed asbestos only on support columns above a suspended ceiling, and in pipe and boiler insulation.

This chapter deals with three issues. The next section examines the effect on building life of a requirement that asbestos be removed prior to demolition of the building. The data from Case 1 are used, because the cost of removal is greater in relation to the building's floor space for this case than for any other case, which produces the maximum impact upon building life. The following section examines the cost-effectiveness of asbestos removal, calculating the social cost of removing asbestos divided by the reduction in risk measured in life-years saved. Four asbestos removal policies are considered for each case: no removal, removal just prior to demolition, removal at renovation, and removal now. The final section examines the financial impact upon the building owner of a rule making him liable for

Table 6-1. Summary of Case Study Building Characteristics

| Building characteristics | Building type | | |
	Office (case 1)	School (case 2)	Office (case 3)
Floor area (000 sq. ft.)	66	26[a]	28
Year built	1959	N.A.	1963
Remaining life (yrs.)	50[a]	30	30[a]
Occupants	263	349[b]	68
		26[c]	
Asbestos			
Type	spray[a]	spray	spray
Area (000 sq. ft.)	66	5.76	N.A.
Removal cost ($000)	N.A.	N.A.	11
Airborne asbestos concentration (f/cc)	.001[a]	.001[d]	.0001[a]

Note: N.A. = not available.
[a] Assumed.
[b] Students.
[c] Staff.
[d] Occupants are exposed 0.63 of a year, so the effective occupant exposure is 0.00063 f/cc.

a specified sum for any asbestos-related death resulting from exposure to asbestos in his building. This calculation is performed only for the privately owned buildings.

The Effect of Asbestos Removal Policies on Building Life

A building owner, who determines the time to demolish his building based upon the net revenue stream from the building and the opportunity cost of the site, will postpone demolition when the cost of demolition is increased by the necessity of removing asbestos or any hazardous material prior to demolition. This principle was established by the timing of demolition analysis in chapter 5. The magnitude of the delay can only be determined by inserting quantitative values in equations (3) and (5) of appendix A. Here we use the data from Case 1 as a starting point, since the cost of removal per square foot of floor space or per dollar of current revenue is higher in this case than in most others, therefore the effect on building life should be greatest in Case 1.

The application of equations (3) and (5) requires data on annual costs and revenues, both present and future. The data base for Case 1 includes both current costs and revenue, and projections of these amounts for ten years in the future. The owner's recent acquisition of this building led him to perform economic calculations which we could use for this analysis. In real terms, revenue is projected to decline by 1.875 percent per year, while costs are projected to remain constant. The site value was estimated from vacant property sales in the area, but no estimate of its future appreciation was available, nor was this part of the owner's own calculations. A 3 percent growth rate for the site value was assumed, on the grounds that the return to holding vacant land must be small, since so little remains in the neighborhood.[1] The owner's cost of capital appears to be about 6 percent in real terms. These data and assumptions are shown in table 6-2.

[1]The owner believes that demolition results not from a building becoming old, but from a profitable opportunity to increase the density of development on the site. This particular building exhausts the allowable density, so the site value would increase only in response to increases in demand for floor space in the area, which should raise rents, or in response to the possibility of rezoning to increase density, something that is very hard to predict far in the future.

Table 6-2. The Effect of Asbestos Removal Costs on Building Life: Case 1 (Office Building)

| | Panel A | | | |
Assumption	Assumed removal cost ($ million)	Life (years)	Building value ($ million)	∂ Life ÷ ∂ removal cost (years/$ million)
No asbestos	0.00	36.3	12.14	2.65
Low average removal cost[a]	0.218	36.9	12.11	2.65
High average removal cost[a]	0.509	37.7	12.08	2.70
Arbitrary high removal cost	2.00	41.6		

Panel B. Sensitivity analysis for removal cost = $218,000

Revenue growth = −0.875%	51.5
Cost growth = 1.0%	31.2
Site growth = 2.0%	38.3
Discount rate = 5.0%	42.5

Note: Data	Current value ($ million)	Growth rate (%/yr.)
Revenue	1.35	−1.875
Cost	0.355	0.00
Site value	3.67	3.00
Discount rate		6.00

[a] From table 4-1.

The calculated optimal life of the building is surprisingly insensitive to the need to remove asbestos before demolition (see panel A, table 6-2). In the absence of any removal cost, the optimal life is 36.3 years. Imposing a low removal cost raises this to 36.9 years, while a high removal cost extends it to 37.7 years. This insensitivity is caused in part by the removal cost amounting to only 5 percent of the current value of the building before discounting, even when the high cost estimate is used. Yet it would be unusual for the cost of removal at demolition, when costs are lower than they are if the building is to be reoccupied, to rise much higher than this, since this building was assumed to have asbestos insulation equal in area to the building's floor space. Raising the removal cost to $2 million only increases the optimal life to 41.6 years. The calculated value of equation (5), appendix A, the derivative of life with respect to removal cost, yields values generally in the vicinity of 2.6 years per million dollars, where one million dollars is about 10 percent of the current value of the building.

The sensitivity of building life to changes in parameters other than the cost of removal is shown in panel B of table 6-2. Increasing the revenue growth rate (that is, making the rate less negative) by one percentage point increases the building life by 15 years, while increasing the cost growth rate reduces life by 5.7 years. Decreasing the growth rate of the site value by one percentage point raises life by 1.4 years. Decreasing the owner's discount rate by one percentage point increases the building life by 5.6 years. These growth rates are subject to considerable uncertainty, yet two of them have a significant effect on building life. This suggests that one should not place too much weight on the prediction of a particular building life from these data. On the other hand, the marginal effect of asbestos removal costs on building life is largely unaffected by changes in the other parameters, so this analysis may be more robust than the basic estimate of the economic life of the building.

Analysis of the elasticity of building life with respect to the three key financial variables produced elasticities of 0.87 with respect to revenue, −0.43 with respect to costs, and −0.38 with respect to site value. As with changes in the growth rates, revenue has the greatest impact, followed by costs and site value. Since the cost of asbestos removal is small relative to the present value of the three variables just listed, changes in its magnitude have a correspondingly small effect on building life.

Since the asbestos removal cost of the building analyzed here is larger relative to the building size and value than for the other buildings analyzed, we conclude that the need to remove asbestos prior to building demolition is not likely to have a large effect on the life of most buildings. Only when the cost of removal is much larger relative to revenues or building values than is the case here will a substantial effect on building life be felt. There may be some situations in which a substantial postponement of building demolition may be expected. However, while ignoring changes in building life, it seems likely that one may analyze the effect of alternative asbestos control policies on building value, or on the behavior of the building owner, without introducing serious error into the analysis.

This analysis has ignored the impact of taxes and depreciation on the determination of building life. As indicated in chapter 5, in the section on material control costs and timing of demolition, the corporate income tax would reduce the effect of asbestos removal costs on the economic life of a building. Thus the effects found here overstate those that should be found in actual practice.

Cost-Effectiveness of Asbestos Removal

The cost-effectiveness of asbestos removal is determined using the social cost-effectiveness model developed in chapter 5. This model is applied to each of the cases and to variations on the cases to determine the sensitivity of the results to the major parameter values. For each case, four scenarios are compared. In the BASE Scenario, the asbestos remains in the building for its projected life. One renovation is assumed to take place, during which the asbestos is disturbed but not removed, and the renovation workers are exposed to airborne asbestos. The asbestos is disturbed again when the building is demolished, and the demolition workers are exposed to high concentrations of airborne asbestos. The health model predicts the risks of death to occupants, building workers, renovation workers, and demolition workers.

In the DEMO scenario, the asbestos is removed just prior to demolition of the building. Building workers and occupants are exposed for the life of the building. Renovation workers are exposed at the time of renovation, and removal workers are exposed at the time of removal. Only demolition workers are not exposed. While the asbestos is in the building, there is an annual cost for maintenance and custodial control programs to minimize the risk of exposure for building workers and occupants. In the absence of data regarding the expected frequency with which the asbestos might be disturbed such that worker protection precautions must be undertaken, a fixed cost per square foot per year is assumed for the control program.

In the RENO scenario, the asbestos is removed prior to renovation of the building. Building workers and occupants are exposed to asbestos in the air until the renovation, and removal workers are exposed. Renovation may occur now or at some later year specified in the scenario data. In the NOW scenario, the asbestos is removed at once, incurring a cost in year one. There is some exposure for removal workers, but none for anyone else.

The present value of the cost of controlling the asbestos is calculated for each scenario, using the social discount rate. Generally both a low and a high cost estimate are available. The cost of removal is treated as a capital investment and multiplied by the shadow price of capital to determine the social cost of that investment. The probable mortality arising in each year is determined from the various exposure levels. All mortality is discounted to the present using the same social discount rate that is applied to costs. The cost-effectiveness of a scenario is computed as the difference in the present value of the costs of that scenario and the next most costly scenario divided by the difference in the present value of the predicted mortality. The costs of the scenarios are such that DEMO is compared to BASE, RENO is compared to DEMO, and NOW is compared to RENO. Cost-effectiveness is expressed as the present value of the cost per life-year saved.

While the calculated cost-effectiveness of a scenario is not by itself a proper basis for accepting or rejecting it, it is useful to compare the cost-effectiveness of various actions with each other. Such a comparison would be useful if one wished, for example, to maximize the reduction in risks achieved by some set of actions, subject to a budget constraint. The solution to this problem would involve choosing those actions with the lowest cost per unit of risk reduction. Appendix C presents evidence on the willingness of individuals to pay to avoid workplace risks and on the cost-effectiveness of a very selected set of public policies involving public expenditure or regulation of the private sector. The range of costs is striking, and the selection of the examples is not necessarily representative. Still, it appears unusual for individuals or the public sector to choose risk-reduction policies that imply a cost in excess of $1 million dollars per life-year saved, for a variety of risks. This figure will be referred to in the discussion that follows, not as a boundary between the reasonable and the unreasonable, but as a reference point for the upper range of the set of estimates presented in the appendix.

Case Studies and Sensitivity Analysis

Case 1.   Since there are no exposure data for this office building, an exposure of 0.001 f/cc is assumed. Chapter 2 suggests that this is a high estimate of exposure in an asbestos-containing building. The renovation is assumed to occur in the fifteenth year, and if the asbestos is not removed then, the renovation workers are assumed to experience some asbestos exposure. The present value of the risk of asbestos-related mortality for occupants, building workers, renovation workers, and demolition workers in the BASE scenario is 0.0208 life-years (see table 6-3). This risk is reduced significantly if removal occurs at renovation. It is barely reduced by removal at once, because the elimination of exposure for occupants and building workers is largely offset by the immediate exposure of the removal workers.

The present value of the cost of removal at demolition is small because of discounting and because removal itself is less expensive at demolition than if the building is to be reoccupied. The cost of the RENO scenario is less than that of NOW because the moving costs are not associated with the asbestos removal. Since RENO occurs sooner, the cost is greater than that of DEMO and discounting has less effect. NOW is the most expensive scenario because the insulation must be replaced, and because moving the occupants and finding them alternative space are costly.

Table 6-3. Cost Effectiveness of Asbestos Removal: Case 1 (Office Building)

	Scenario			
Run	BASE	DEMO	RENO	NOW
Basic run				
Risk (present value, life-years)	0.0208	0.0170	0.0165	0.0197
Cost (present value, $ million)				
Low cost estimate		0.105	0.687	2.11
High cost estimate		0.227	1.45	3.88
Cost-effectiveness ($ million/life-year)				
Low		28.0	1.14	N.A.
High		60.5	2.41	N.A.
Cost-effectiveness ($ million/life)				
Low		310.	12.7	N.A.
High		670.	26.7	N.A.
Discount rate = 0				
Cost-effectiveness ($ million/life-year)				
Low		0.737	0.407	3.63
High		1.59	0.784	5.15
Ignore removal, demolition risks				
Cost-effectiveness ($ million/life-year)				
Low		N.A.	65.8	237.
High		N.A.	138.	403.
Alternate occupant exposure: 0.01 f/cc				
Cost-effectiveness ($ million/life-year)				
Low		28.	7.27	27.9
High		60.5	15.3	47.6

Notes: N.A. means no risk reduction, marginal cost is not defined.

Data:	Number	Exposure (f/cc)
Occupants	263	0.001
Workers	0.2	0.01
Removal	0.91 man-decades	1.0
Demolition	0.50 man-decades	5.0

Removing the asbestos prior to demolition reduces mortality risks at a cost of $28 million to $60.5 million per life-year saved in the basic run (the run using the baseline parameter values). Expressed in terms of lives rather than life-years, the cost is $310 million to $670 million per life saved. In addition to reducing total risks, the DEMO policy shifts risks, causing risks for removal workers, but eliminating risks that are about twice as large for demolition workers. Since removal of asbestos now or at renovation does not reduce discounted risks below the risks of removal at demolition, the cost per life-year saved is not defined.

These basic results change in several ways if some of the assumptions are changed. If future costs and mortality are not discounted, the cost per unit of risk reduction drops for all policies, but their rank ordering is unchanged. Removal NOW does reduce risks, but it is still the least preferred policy. RENO is the most cost-effective of the policies, as before, with a cost per life-year saved of $0.407 million to $0.784 million per life-year saved. If the exposures of workers engaged in renovation and demolition work are assumed to be zero, then removal at demolition yields no health benefits, and removal at renovation or now incurs very high costs of $65 million to $403 million per life-year saved. Finally, assume that the condition of the ACFM in the building is badly deteriorated, or that the material is regularly disturbed, multiplying the exposure of building occupants and of maintenance and custodial workers by a factor of 10, yielding exposures of 0.01 and 0.1 f/cc, respectively. This multiplies the total risk in the BASE and DEMO scenarios by about a factor of 7. Under this assumption, removal at renovation reduces total risk by a considerable amount, and the cost per life-year saved drops to $7.27 million or $15.3 million (see table 6-3). While this is still expensive, it is less costly per life-year saved than removal now or at demolition.

Case 1 shows that all scenarios and all assumptions yield costs that are far above the level of $1 million per life-year except when a discount rate of zero is assumed. Removal now or at demolition always costs more per life-year saved than removal at renovation. Removal now is not preferred to removal at demolition unless we may assume that the exposure of the removal workers themselves is negligible. Even if the asbestos insulation is damaged or is being disturbed, and airborne asbestos fiber levels are on the order of 0.01 f/cc, removal at renovation is more cost-effective than removal now. Under all assumptions, however, in this building any asbestos removal is a very expensive means of reducing the risk of premature mortality.

Case 2. This average school incorporates asbestos insulation that covers an area less than one-fifth the floor area of the school. Despite this modest amount of asbestos insulation, the airborne asbestos fiber level assumed[2] by the EPA is 270 ng/m^3, which may be converted

[2]EPA derived this figure from the Constant study (1983), using the arithmetic average exposure in rooms containing ACFM. A geometric average over *all* rooms, both with and without ACFM is close to 0.002 f/cc.

to about 0.0082 f/cc. In chapter 2 it was noted that subsequent EPA documents present lower estimates of exposure levels in schools. We will therefore assume that the average airborne asbestos fiber level in this school would be 0.001 f/cc. Since students are in school only 7 hours per day, 180 days per year, their annual exposure is 0.63 times that of a full time worker exposed to the same fiber level for 8 hours per day, 250 days per year. The students are assumed to be first exposed at age 7, since not all students in the school will start in the earliest grade, and to be exposed for 10 consecutive years. Custodial and maintenance workers are assumed to be exposed to fiber levels ten times those of the students, or 0.01 f/cc. Nonstudents are assumed to be first exposed at age 35, since only a small proportion of such workers will have their first work experience in a school containing asbestos. The school will have 26 teachers and administrators, also assumed to be exposed to 0.001 f/cc for 63 percent of a full working year. Since there are no moving costs, because removal can occur during summer holidays, the cost of removal now is best estimated by using the calculated cost of removal at renovation.

Table 6-4 shows that the present value of the risk for students, teachers, building workers, renovation workers, and demolition workers in the BASE scenario is 0.0176 life-years, almost 90 percent of which represents risks to students. This risk is barely reduced by removal at demolition, but is greatly reduced, to 0.0017, by removal at renovation, assumed to occur now. The considerable reduction in risk between the DEMO and RENO scenarios occurs because of the high occupant density in the school. While the young age of the students approximately doubles their lifetime risk of premature mortality, compared with a cohort first exposed at age twenty-two, the discounted value of this risk increases by less than 25 percent because the deaths occur farther in the future for those exposed first at age seven. The present value of the cost of removal at demolition is only $21,000 to $47,000, while removal at renovation costs $176,000 to $365,000.

Removal at demolition costs $26.1 million to $59.2 million per life-year saved, while removal at renovation costs $10.3 million to $21.1 million per life-year saved. Removal at renovation is less costly per life-year saved than removal at demolition. This was also true for Case 1. This pattern persists for all runs presented in table 6-4. While even removal at renovation costs more than $10 million per life-year saved, the cost per life-year is lower than the cost of removal at any time in Case 1.

These results, like those of Case 1, are sensitive to several as-sumptions. If future costs and mortality are not discounted, removal

Table 6-4. Cost-Effectiveness of Asbestos Removal:
Case 2 (School Building)

	Scenario			
Run	BASE	DEMO	RENO	NOW
Basic run				
Risk (present value, life-years)	0.0176	0.017	0.0017	0.0017
Cost (present value $ million)				
Low cost estimate		0.021	0.176	0.357
High cost estimate		0.047	0.365	0.93
Cost-effectiveness ($ million/life-year)				
Low		26.1	10.3	N.A.
High		59.2	21.1	N.A.
Cost-effectiveness ($ million/life)				
Low		258.	146.	N.A.
High		585.	265.	N.A.
Discount rate = 0				
Cost-effectiveness ($ million/life-year)				
Low		1.32	0.44	N.A.
High		3.0	0.91	N.A.
Ignore removal, demolition risks				
Cost-effectiveness ($ million/life-year)				
Low		N.A.	10.9	N.A.
High		N.A.	22.4	N.A.
Alternate occupant exposure: 0.00517 f/cc				
Cost-effectiveness ($ million/life-year)				
Low		26.0	1.17	N.A.
High		59.2	2.4	N.A.

Notes: N.A. means no risk reduction; marginal cost is not defined.

Data:	*Number*	*Exposure*
Occupants	349 + 26	0.00063 f/cc (0.001 for 0.63 years)
Workers	3.4	0.01 f/cc
Removal	0.079 man-decades	1.0 f/cc
Demolition	0.043 man-decades	5.0 f/cc

at renovation or demolition costs only 5 percent as much per life-year saved compared to the discounted runs. Ignoring exposures at removal and demolition predictably renders removal at demolition unproductive for reducing risk and barely alters the cost-effectiveness of removal at renovation. Assuming that the exposure of occupants averages 0.00517 f/cc rather than 0.00063 f/cc and that worker exposure averages 0.082 f/cc rather than 0.01 divides the cost of removal at renovation per unit of risk reduction by almost a factor of ten, reflecting the reduced reduction in risk for occupants and workers. See table 6-4.

Thus if school exposures were in fact as high as the EPA study assumed, the cost-effectiveness of asbestos removal in schools would be far better than that of asbestos removal in other buildings. As indicated earlier, the EPA does not support these high exposure estimates as representative of typical schools.

Case 2 suggests that several factors may cause the cost-effectiveness of asbestos removal in schools to differ from that in office buildings. First, there are generally more students per square foot of floor space in a school than there are office workers per square foot of office space. This means that any reduction in airborne asbestos levels creates more health benefits in a school than in an office building. Second, because school buildings are often vacant in the summer and during other vacation periods, it is often possible to conduct asbestos removal work without disrupting the work of building occupants. In an office building, the disruption of the occupants can be as costly as the removal work itself. A third factor that distinguishes Cases 1 and 2, but may not necessarily distinguish offices and schools, is that the school in Case 2 contains a small amount of asbestos, resulting in small exposures of removal workers, while the office in Case 1 contained a large amount of asbestos, causing concomitantly large exposures of removal workers. All three factors combine to make removal at renovation in Case 2 more cost-effective than removal at demolition.

Case 3. This building has sprayed asbestos only on the tops of columns above a dropped ceiling. Air sampling resulted in no detectable airborne asbestos, but no information is available on the type of analysis performed. Assuming that TEM analysis was used, an airborne level of 0.0001 f/cc is assumed, since this is a typical detection level for that measurment method. A life of thirty years is assumed for the building, and it is assumed that renovation will take place in ten years.

The present values of the risks and costs are shown in table 6-5. Discounted risks are lowest for removal at demolition, because occupant exposures are so low. Removal at any other time raises risks, and is therefore not cost-effective. Removal at demolition costs $2.38 million per life-year saved, considerably less than in the previous two cases. This reflects a very low cost for removal, based upon a contractor's estimate.

If future costs and mortality are not discounted, removal now or at renovation becomes cost-effective, but far more costly, per life-year saved, than removal at demolition. Unlike Case 1, the absence of discounting cannot overcome the very low occupant risks pre-

Table 6-5. Cost-Effectiveness of Asbestos Removal: Case 3 (Professional Building)

Run	BASE	DEMO	RENO	NOW
		Scenario		
Basic run				
Risk (present value, life-years)	0.00449	0.00053	0.00058	0.00060
Cost (present value, $ million)				
Low		0.009	0.087	0.40
High		0.009	0.140	0.63
Cost-effectiveness ($ million/life-year)				
Low		2.38	N.A.	N.A.
High		2.38	N.A.	N.A.
Cost-effectiveness ($ million/life)				
Low		26.4	N.A.	N.A.
High		26.4	N.A.	N.A.
Discount rate = 0				
Cost-effectiveness ($ million/life-year)				
Low		0.233	28.3	159.
High		0.233	54.2	243.
Ignore removal, demolition risks				
Cost-effectiveness ($ million/life-year)				
Low		N.A.	419.	1777.
High		N.A.	735.	2742.
Alternate occupant exposure: 0.001 f/cc				
Cost-effectiveness ($ million/life-year)				
Low		2.38	55.6	230.
High		2.38	97.6	355.

Notes: N.A. means no risk reduction, marginal cost is not defined.

Data	Number	Exposure
Occupants	68	0.0001 f/cc
Workers	0.1	.01 f/cc
Removal	0.0275 man-decades	1.0 f/cc
Demolition	0.14 man-decades	5.0 f/cc

sented by this building. If the removal and demolition risks are ignored, removal now or at demolition becomes cost-effective, but ranges in cost from $419 million to $2,742 million per life-year. Assuming occupant exposures to be ten times greater, or 0.001 f/cc, yields costs for removal at renovation of $55.6 million to $97.6 million per life-year, still more than ten times the cost of removal at demolition (see table 6-5).

Case 3 shows that if the amount of asbestos is small and the exposure of occupants is at a very low level, removal at demolition is the most attractive strategy, and even then it costs more than $1 million per life-year saved. Removal now is invariably enormously expensive per life-year saved.

Private Incentives for Hazard Control

A building owner may be required by regulations or by demands from his tenants or maintenance employees to undertake control actions to reduce risks presented by asbestos in a building. It is also possible to create incentives to undertake control actions if future asbestos-related disease will impose costs upon the building owner. Suppose, entirely hypothetically, that any asbestos-related cancer death resulting from exposure to airborne asbestos in a building would result in liability by the building owner, at the time of the death. Suppose that the magnitude of this liability were V dollars per life-year lost. The building owner would consider the future stream of potential liability costs, as well as the cost of control, in deciding what policy would maximize profits, and thus maximize the present value of the building.

The analysis that follows employs the cash flow model of the firm presented in chapter 5 to determine the present value of the after-tax cash flow of the building owner. The liability, V, is assumed to be $1 million per life-year lost, almost certainly an overestimate of what a court would award today, even if liability itself could be established (Dewees, 1985, p. 20). Case 2 is not analyzed because it is a school and not only would liability be different from that of a private building owner, but the financial model of the private firm is not applicable.

Table 6-6 shows the present value of the net cash flow associated with asbestos control when liability is set at $1 million per life-year. With this liability rate, in no run is any scenario less costly than the BASE Scenario, in which no control action is taken. In the basic runs, even removal at demolition is five to ten times as costly as the BASE Scenario. Removal at renovation is about ten times more costly still. Ignoring removal and demolition exposures leaves the results virtually unchanged. Increasing occupant exposures increases the cost of the BASE and DEMO policies, but leaves the ranking of the policies unchanged. Even with an unrealistically high liability if $1 million per life-year, an owner would have no economic incentive to remove the asbestos at any time. Only when the liability is increased to

Table 6-6. Private Incentives for Asbestos Control: High Cost Estimate Only

Scenario	Present value of cash flow ($ million)			
	BASE	DEMO	RENO	NOW
Basic Run				
Case 1	−.0022	−.028	−.21	−.69
Case 3	−.00039	−.0015	−.018	−.087
Ignore removal, demolition risk				
Case 1	−.0020	−.027	−.21	−.69
Case 3	−.0052	−.0014	−.018	−.086
Alternate occupant exposure				
Case 1 (0.01)	−.020	−.045	−.22	−.690
Case 3 (0.001)	−.00080	−.0019	−.018	−.087
Liability = $100,000,000/life-year				
Case 1	−.23	−.23	−.43	−1.01
Case 3	−.039	−.0079	−.026	−.096

$100,000,000 per life-year, in the last two lines of table 6-6, is removal at demolition as attractive as no removal at all.

Two factors contribute to the inability of this incentive rule to create an effective incentive to remove asbestos from a building. First, while $1 million per life-year is a high liability rate, the social cost-effectiveness analysis in the preceding section generally found removal action to cost still more. Asbestos control in buildings simply costs far more than most courts have been willing to award for loss of life. Second, the model of the firm employs a discount rate greater than that used in the social cost-effectiveness analysis. With most of the loss of life occurring decades after any control investment is made, the liability for disease is highly discounted.

The reduction in the discounted risk of loss of life occurring when asbestos removal actions are undertaken have been quite small in all cases. The social cost per life-year saved by control actions is high. It should not therefore be surprising that even high award levels under a tort liability regime fail to create effective incentives to undertake control actions. This is true even when we assume, as we have here, that the building owner has perfect information about the asbestos in his building, that he can forecast the future health risks that this asbestos raises, that every fatality caused by this asbestos will result in a successful lawsuit against the owner of the building, and that the award to the decendent's plaintiff will amount to $1 million per life-year lost, or more than $10 million per life lost.

In fact, since these assumptions are highly optimistic, the actual incentive created by a tort liability regime might be even smaller than estimated here.

On the other hand, the liability of the building owner might not in fact be limited to liability for disease actually caused by asbestos exposure in the building. A population consisting of half smokers and half nonsmokers that is exposed in a building to airborne asbestos fiber concentrations of 0.001 f/cc for ten years may expect premature mortality risks from lung cancer of 75,000 per million attributable to smoking alone, and sixteen per million from the asbestos exposure. If all persons who worked in the building and later contracted lung cancer were to succeed in convincing the courts that their disease was related to the asbestos exposure in the building, the owner's liability would be thousands of times greater than that based upon the predictions of excess lung cancer caused by the asbestos exposure alone. It is difficult to predict whether courts will ever allow recoveries by building occupants for asbestos-related lung cancer, and if so, what proportion of total lung cancer victims might win such lawsuits. This renders tort liability an enormously unreliable mechanism for inducing building owners to deal with asbestos in a responsible manner.

It is true that this analysis deals primarily with asbestos material that is in good condition and is not being disturbed, so that the airborne asbestos fiber levels are quite low. However, the variation on Case 1 considered an airborne asbestos fiber level of 0.01 f/cc, which might be caused by asbestos material that is damaged or being disturbed. Even in this variation, the least cost alternative for the firm is not to control at all. Given the facts of these cases, one would have to assume airborne asbestos fiber levels that were much greater still, exceeding 0.1 f/cc, before tort liability of the type examined here would cause removal now or at renovation to be less costly than leaving the material in the building. Alternatively, tort liability might accelerate the time of removal if the cost of precautions required while the asbestos remains in the building is greater than that assumed here.

REFERENCES

Constant, Paul C., Fred J. Bergman, Gaylord B.R. Atkinson, and coauthors. 1983. *Airborne Asbestos Levels in Schools*, EPA 560/5-83-003 (Washington, D.C., U.S. Environmental Protection Agency, Office of Toxic Substances).

Dewees, Donald N. 1986. "Economic Incentives for Controlling Industrial Disease: The Asbestos Case," *Journal of Legal Studies* vol. 15, no. 2 (June) pp. 289–320.

Seven

CONCLUSIONS

No one can doubt the seriousness of the health hazards posed by the inhalation of high concentrations of asbestos fibers over a substantial period of time. The tragic legacy of the exposure of industrial workers and insulation workers to high asbestos concentrations from World War II to the late 1960s has been documented and analyzed in a dozen or more epidemiological studies. These studies clearly establish the linkage between worker exposures to asbestos and the development of asbestosis, lung cancer, and in some cases mesothelioma. A crucial factor established by these studies is that the high disease incidence is generally associated with exposures to high dust concentrations over a period of years. Asbestos insulation workers who installed the insulation that is the subject of current concern were exposed, usually for many years, to airborne asbestos fiber levels that have been estimated to average between 5 and 10 f/cc. In short, the cumulative exposure of the workers in these cohorts has been quite large.

By contrast, the exposure of building occupants to airborne asbestos fiber concentrations is extremely small according to the available public data. The available evidence, reviewed in chapter 2, supports the conclusion that occupants of buildings containing friable asbestos insulation that is in good condition and not being disturbed will in general be exposed to average concentrations of airborne asbestos less than 0.001 f/cc. While in some buildings exposures may exceed 0.01 f/cc, there is no evidence that typical building exposures exceed one one-thousandth f/cc. These typical exposures are between one one-thousandth and one ten-thousandth of the exposure intensities which the insulation workers experienced.

The exposures of building custodial, maintenance, renovation, and asbestos removal workers are more difficult to estimate than those of building occupants, but are likely to be higher in situations where this work is in the vicinity of friable asbestos-containing insulation. The evidence reviewed in chapter 2 indicates that maintenance or renovation work that disturbs friable asbestos-containing insulation. will cause airborne fiber levels in the breathing zone of the workers ranging from 0.1 to 0.5 f/cc and sometimes much higher. Removal of such insulation may cause airborne fiber levels ranging from 0.3 f/cc for careful work using wet methods to more than 10 f/cc for dry removal. The 10 f/cc level is within the range of estimates of the average exposures to which insulation workers were exposed in earlier decades. Clearly, fiber concentrations of 10 f/cc should be a matter of serious concern unless care is taken to reduce actual worker exposures to levels far below the airborne concentration.

Predicting the health effects of current building exposures to asbestos involves predicting lung cancer and mesothelioma risks. There should not be a significant risk of asbestosis so long as the long term average exposure of individuals is well below one f/cc. This will certainly be the case for building occupants, and should be the case for building workers so long as exposures during maintenance, renovation, and removal activities are carefully controlled. Predicting the incidence of lung cancer and mesothelioma is difficult because exposures in buildings are far less intense than the past exposures of workers whose health experience forms the basis for understanding the dose-response function. Any estimate of the disease rate that might result from low level exposures is thus subject to considerable uncertainty. Despite these difficulties, models of the dose-response function for lung cancer and mesothelioma have been developed. The health risk models reviewed in chapter 3 all employ a relationship between the cancer risk and the airborne asbestos fiber concentration that is linear through the origin, so that risks are directly proportional to the airborne fiber level. The relationship between cancer risks and the age of exposure, smoking habit, and duration of exposure is more complex, but there is great similarity among the models as to how these are treated. The result is that most models yield similar predictions of health risks for building exposures.

Assuming that cancer risks are proportional to exposure levels, and given that typical exposures of building occupants are one one-thousandth to one ten-thousandth as great as those of insulation workers in past decades, the health risks of building occupants are in the range of one one-thousandth to one ten-thousandth those of the insulation workers. The models of chapter 3 predict that if a group of one million individuals were exposed to 0.001 f/cc in a building for ten years beginning at age twenty-two, by the time all one million persons had died of various causes, one could expect to find sixteen deaths from asbestos-related diseases. Comparing this risk to average risks faced by the American public, it is less than one one-thousandth the risk of dying from an auto accident, one one-hundredth the risk of drowning, and one five-hundredth the risk of being a victim of suicide or homicide, or of cancer from all causes. A commuter who drove 5 miles each way for 10 years to a building with airborne asbestos levels of 0.001 f/cc would be twenty times more likely to die in an auto accident while commuting than to die from asbestos disease decades later. Thus the very low levels of asbestos fiber exposure for building occupants lead to low levels of health risks. The Ontario Royal Commission on Asbestos concluded that the health risks to occupants of buildings in which asbestos insulation was in good condition, not disturbed, and not falling onto building surfaces, was insignificant. Nothing has happened in the two years since that report was issued to cause me to disagree today with that assessment.

The analysis in chapters 5 and 6 shows that a requirement that asbestos be removed prior to building demolition will tend to prolong the economic life of the building by adding to the cost of demolition. The magnitude of this effect increases with the cost of removal in relation to the value of the building itself. In the cases examined here, the removal cost is relatively small, so the magnitude of this effect is quite small, amounting to a maximum of only a few years in a building with a life of thirty-six years. It is possible, however, that a building with asbestos that is costly to remove, and for which the opportunity cost of the site is modest, may have its life significantly prolonged by the anticipated cost of removal at demolition.

The low health risk to building occupants when the asbestos is in good condition and undisturbed, combined with the often high cost of removing asbestos insulation from a building, causes the cost of asbestos removal programs to be quite high when measured against the number of life-years saved.

Most of the cases and variations on them studied here yield discounted costs greatly in excess of $1 million dollars per life-year saved. Even if future health effects (and cost) are not discounted, costs are often above $1 million per life-year saved, and over $10 million per life saved. In comparison with other public programs, far more lives can be saved with the same expenditure by investing in highway safety, in a variety of occupational health programs, and in providing medical services (see app. C). School buildings often have more occupants per square foot than office buildings, and the summer vacation allows asbestos control work that does not interfere with normal business; therefore the cost-effectiveness of asbestos removal in schools is often considerably better than in office buildings. Still, the costs examined here generally exceed $1 million dollars per life-year saved even in school buildings.

The U.S. EPA as well as the Ontario Ministry of Labour requires that friable asbestos be removed prior to demolition of the building. The question then is not *whether* to remove the asbestos, but *when*. The calculations of chapter 6, which assume that the cost of ensuring the safety of building occupants and workers is low, show that removal at demolition is far less expensive than removal at renovation, which in turn is far less expensive than removal now. On a cost basis alone, the clear choice is to defer removal. Computing the cost of removal per unit of risk reduction for all who may be exposed to asbestos in the building reveals that the most cost-effective time for removal is at renovation. Removal at demolition is less cost-effective, and removal now is least cost-effective. These results, of course, are valid for the analyzed cases, and do not prove that in other situations other policies may not dominate. For example, if the frequency of disturbance of the material is high, the precautions needed to ensure worker safety might be sufficiently expensive that removal now is less expensive than deferring removal until renovation or demolition. This supports the recommendation of the Royal Commission that immediate removal not be mandated, but that the owner should be allowed to choose between removal at demolition or at some earlier time.

Despite the small set of cases examined here, considerable variability is found among buildings in both the exposure of occupants and the cost of controlling or removing asbestos. This variability means that simple universal programs requiring the same actions in all buildings are likely to be expensive and inefficient. The results of this study fail to provide support for a crash program to remove all asbestos at once, not only because this would on average be a very expensive means of reducing health risks to the population, but also because in some cases the risk reduction will be very small. Some buildings will warrant immediate action, others will not. While general principles can be developed for determining when corrective action is appropriate in a building, the facts of each building must be examined individually to apply such principles.

Since workers who remove asbestos from a building may experience substantial exposures to asbestos, any asbestos removal program necessarily involves shifting risks from one group of people to another. In some cases, a large reduction in risk for building occupants and workers may be achieved with a small increase in risk for asbestos workers. In other cases, particularly when the risk presented by the building in its present state is small, the increased risk for removal workers may actually exceed the reduction in risk for building occupants and workers. The cases provide several examples in which removal now or at renovation poses greater aggregate risks than removal at demolition.

Even when asbestos removal reduces aggregate health risks, the cost of this risk reduction in millions of dollars per life or life-year is far above amounts that courts typically award in wrongful death actions. This means that even if building owners were liable for tort damages for all deaths resulting from the asbestos in a building, there would be in general no substantial incentive for the owner to remove the asbestos now. Only if current exposures were very high or if the damages awarded far greater than those assumed here would a substantial incentive emerge to control asbestos.

This ineffectiveness should not be viewed entirely as a defect, since it reflects in part the lack of economic justification for immediate asbestos removal in many buildings.

There are major uncertainties in calculating the liability that building owners might expect as a result of the presence of friable asbestos-containing material in a building. The discount rate that they apply to calculations of future liability may differ among them. The ability to estimate current asbestos fiber levels and to forecast future asbestos-related disease incidence will vary greatly. Predictions of the amount that courts might award to successful plaintiffs in personal injury cases could differ by several orders of magnitude among owners and their lawyers. All these uncertainties would be dwarfed by the unknown proportion of legitimate claims that might succeed in court. If, for example, all persons who worked in a building and later contracted lung cancer were to succeed in convincing the courts that their disease was related to the asbestos exposure in the building, the owner's liability would be thousands of times greater than that based upon the predictions of excess lung cancer caused by asbestos exposure in buildings. This uncertainty is illustrated by the difficulties that asbestos-removal firms faced in 1985 as insurance companies refused to insure against any losses related to asbestos (or other environmental hazards), because they have no basis for estimating the risks involved. For those who believe that asbestos removal in buildings deserves high priority, the imposition of tort liability on building owners appears to be an unpredictable instrument for achieving that goal.

The analysis performed here suggests that crash programs to remove all friable asbestos-containing material from buildings are a very expensive means of protecting the health of building occupants. Consider a building in which the material is in good condition, is not falling onto building surfaces, and is not being disturbed. Postponement of asbestos removal until maintenance, renovation, or demolition of the building would disturb the asbestos-containing material costs considerably less than immediate removal and imposes little added risk on building occupants. On the other hand, the intensity of exposure to airborne asbestos fibers may be far greater for building maintenance and renovation workers than for building occupants. Therefore the exposures of these workers should be a central consideration in choosing among asbestos control strategies. Finally, the cost of asbestos control strategies must be considered in choosing among those strategies. Building owners, including public agencies, have varied opportunities to reduce risks for building occupants at some cost. Money spent on asbestos control programs may reduce spending on other programs which might yield greater health or safety benefits.

There is a need for substantial research efforts in several areas related to asbestos control programs. More reported data are needed on the airborne asbestos fiber concentrations in buildings in order to reduce the present uncertainty over likely concentrations in various types of buildings. The cost of operations and maintenance programs that minimize exposure risks to occupants and workers alike must be studied to fill what is now a void in the literature. Finally, it would be helpful to learn more about the relative risks presented by different asbestos types in building insulation, and about the shape of the dose-response function at the low fiber levels prevalent in buildings. As better data are developed in each of the above areas, more accurate and comprehensive economic calculations of the type presented in this study will become feasible.

APPENDIX A
MODEL OF REMOVAL COSTS AND BUILDING LIFE

Assume that revenues and costs associated with an existing building and the value of the site may grow over time, but that costs grow more rapidly than revenues. To simplify the problem, assume specific forms for the three functions: $R_t = Re^{at}$, $C_t = Ce^{bt}$, and the value of the clear site $VS_t = VSe^{gt}$. Assume that in general the firm's discount rate exceeds the growth rate of revenue or cost; that is: $r > a,b$. Assume further that as the building ages, costs rise at least as rapidly as revenues; $b \geqq a \geqq 0$. Over time, annual profits will fall until at some time it is more profitable to demolish and replace the building than to retain it. The cost of removing the material just prior to demolition is A. The building owner's discount rate, derived from other investment opportunities, is r. Initially we will omit taxes and depreciation to simplify the mathematics.

The owner wishes to choose the time of demolition T to maximize:

$$V(T) = \int_0^T e^{-rt}[Re^{at} - Ce^{bt}]dt + e^{-rT}[VSe^{gT} - A] \qquad (1)$$

where the first expression is the present value of the profit earned until demolition, and the second term is the present value of the building site net of demolition costs. The first order condition for maximizing this expression, using Liebnitz' rule, is:

$$V'(T) = e^{-(r-a)T}[R - Ce^{(b-a)T}] - (r - g)VSe^{-(r-g)T} + rAe^{-rT} = 0 \qquad (2)$$

which yields, after canceling e^{-rt}:

$$e^{aT}[R - Ce^{(b-a)T}] = (r - g)VSe^{gT} - rA \qquad (3)$$

This cannot be explicitly solved for T. However we can take the partial differential with respect to the cost of material removal A:

$$-C(b - a)(\partial T/\partial A)e^{(b - a)T}$$
$$= -raA(\partial T/\partial A)e^{-aT} + (r - g)VS(g - a)e^{(g - a)T}\partial T/\partial A - re^{-aT} \tag{4}$$

Solving this for $\partial T/\partial A$ yields:

$$\partial T/\partial A \tag{5}$$
$$= re^{-aT}/[C(b - a)e^{(b - a)T} + arAe^{-aT} + (r - g)(g - a)VSe^{(g - a)T}]$$

The first two terms in the denominator are positive given that $b \geq a$ as assumed. The sign of the last term is positive if $g > a$, and $r > g$, or if $g < a$ and $r < g$, and negative otherwise. If g is less than a, the last term should be smaller than the first term, since the exponent will be negative. Thus the entire expression is positive under most assumptions; that is, T is likely to be an increasing function of A.

If we simplify the functions by assuming that revenue and the value of the site are constant, that is, $a = g = 0$, then equation (3) can be solved for T:

$$T = (1/b) \ln [(R - r(VS - A))/C] \tag{6}$$

Equation (6) also shows that T is an increasing function of A so long as the argument of the log is positive. Thus increasing the cost of material removal postpones the optimal time of demolition of the building. Alternatively, we assume that g is positive, but that $a = b = 0$. Then equation (3) can be solved for T:

$$T = (1/g) \ln [(R - C + rA)/(r - g)VS] \tag{7}$$

The log is only defined if $(R - C + rA)/(r - g) > 0$. In this case, T is still an increasing function of A.

Corporate income taxes may be included in the function to be maximized by the firm, and in the derivation of the firm's discount rate, rewriting equation (1) as:

$$V(T) = \int_0^T e^{-rw^t}(Re^{at} - Ce^{bt} - IT_t)dt + e^{-rwT}(VSe^{gT} - A) \tag{1a}$$

where IT_t is the income tax resulting from the cash flow in year t and r_w is the after-tax weighted cost of capital for the firm, in which the cost of equity and of debt are weighted by the proportion of capital raised from each source. Because corporate income tax exemptions effectively subsidize expenses, the after-tax cost of removing the hazardous material is less than the cost if taxes are ignored. Thus a requirement to remove the hazardous material should retard demolition less in a world with taxes than in a world without taxes.

APPENDIX B
MODEL OF THE FIRM

We assume that the objective of the firm is to maximize the present discounted value of the expected after-tax cash flow of the firm. This objective is shown in equation (1):

$$\text{Max } V_0 = \sum_{t=0}^{\infty} \frac{E\{CF_t\}}{(1 + r_w)^t} \tag{1}$$

where E = expected value
CF_t = after-tax cash flow in year t
r_w = weighted after-tax cost of capital

In equation (1), the firm discounts future cash flows using the weighted after-tax cost of capital as the discount rate. This cost of capital is defined in equation (2):

$$r_w = \alpha r_e + (1 - \tau)(1 - \alpha)r_b \tag{2}$$

where α = fraction of capital in equity
r_e = cost of equity
τ = marginal corporate income tax rate
r_b = cost of debt

The cost of equity capital for the firm may be determined using the capital asset pricing model (CAPM) which predicts the rate of return on a particular stock as r_e, defined in equation (3):

$$r_e = r_f + \beta(r_m - r_f) \tag{3}$$

where r_f = risk-free rate of return
β = the beta of the firm: the correlation of the firm's stock value to variations in the stock market as a whole
r_m = the market rate of return

The risk-free rate, r_f may be represented by the short-term return on Treasury Bills, and r_m by the return on a market portfolio.

The cash flow in equation (1) consists of revenues less expenses, taxes, and capital investment. The taxes in turn are calculated on revenues minus expenses and depreciation. Cash flow may therefore be defined as shown in equation (4):

$$CF_t = (R_t - D'_t - OC'_t)(1 - \tau) + D'_t - K'_t \tag{4}$$

where R_t = revenue from the property
D'_t = depreciation in year t, on the building and on any investment for controlling exposures to the hazardous material
OC'_t = costs in year t, including normal operating costs, costs of dealing with the hazardous materials, and any payment for health injuries
K'_t = capital investment in year t

Depreciation D'_t is determined as some rate times the remaining undepreciated capital stock. The method of calculating depreciation will depend upon the tax rules. Operating costs, OC'_t, include normal operating costs for production, extra costs of dealing with the hazardous material, and any payment for health injuries.

This model of the firm is used to compare the present value of the firm, assuming different policies for controlling hazardous material, different liability regimes, and different assumptions about health effects. If revenues are affected by these policy choices, the change in operating revenue, ΔR_t, appears in the working version of equation (4). Since the cost function is separable, the operating costs that are unrelated to the hazardous materials are the same in all evaluations, and therefore drop out of equation (4), leaving only costs of health protection, wage premiums, and health liabilities in the term labeled OC_t. Investment in the basic building is also the same in all evaluations, as is depreciation on that building, so K_t and D_t are defined to include only investment to deal with the hazardous material. In summary, marginal investments in controlling the hazardous material, which may reduce future liabilities OC_t, may be evaluated by comparing the present value of equation (1) using the cash flow from equation (5):

$$CF_t = (\Delta R_t - D_t - OC_t)(1 - \tau) + D_t - K_t \tag{5}$$

In equation (5), all differences between policies will be a result solely of revenue changes, investments, expenses, and liabilities associated with the hazardous material and its control.

We assume that investment is financed by increasing the financial capital of the firm in an amount just sufficient to cover the investment. In this case, the investment itself would have a negative impact on cash flow, just equaling the payment K_t for the investment. On the firm's balance sheet the liability represented by the financial capital is just offset by the value of the asset—the modified building, represented by K.

What is the beta of investing in a project for controlling the hazardous material? Beta is defined as the correlation of the return on this investment and the return on the stock market as a whole. Beta for the market itself is 1.0, while the beta of an investment that has a return which is uncorrelated with the market is 0. In this model, the benefit of hazardous material control is a reduction in claims for losses due to future disease. The size of this benefit depends upon the effectiveness of the protection, future workers' compensation and tort liability rules, the number of persons exposed, and emerging medical knowledge. The first two of these will have some uncertainty that should be unrelated to other risks of the firm or market. The number of persons exposed should be roughly proportional to the occupancy rate of the building and therefore highly correlated with a major financial risk of the firm—demand for space in the building. This should cause the beta for the project to be less than or equal to the beta for the firm. Turning finally to emerging medical knowledge, this should be uncorrelated with demand for space in the building. Once the material is removed or controlled, the discovery that the material was especially hazardous will not affect the demand for space in the building. If uncertainty about health effects were the primary source of uncertainty for this project, this should cause the beta for this project to be zero, since medical knowledge should be uncorrelated to market risks. In summary, the beta for this project should be between 0 and the beta for the firm. There is nothing about the control of hazardous material to make it especially sensitive to market movements; that is, to give it a beta greater than that of the firm.

APPENDIX C
THE COST OF SAVING LIVES

Job Market Data: The Compensating Wage

One source of evidence on the amount that individuals must be paid to accept increased risks of mortality is the wage premium that must be paid to attract workers to particularly risky jobs. The data in table C-1 represent the amounts that workers demand to accept risks of fatality on the job, as revealed by the compensating wage differential. This is not the amount that a worker would demand to accept certain death tomorrow, but rather the implicit value of life given the additional compensation required to accept a small increase in the probability of death. Each of these studies has examined wage rates in various occupations and attempted to explain variations in those wage rates by using a number of factors including the mortality rate from occupational causes. While studies of the relationship between wages and occupational injuries have produced mixed results, studies of mortality and wages have yielded consistent positive relationships, implying that workers who are exposed to an excess risk of mortality on the job must be compensated for accepting the risk (Rea, 1983). The cause of death is usually an on-the-job accident, so there is no significant time lag between the compensation and the risk of death. The strengths and weaknesses of these studies are discussed in Rea (1983), Bailey (1980), Smith (1979), and Viscusi (1983).

The data in table C-1 show values of life ranging from $140,000 to $4.16 million, in 1967 to 1978 U.S. dollars, or from $0.435 million to $7.3 million 1983 U.S. dollars. Eliminating the studies reporting the highest and lowest estimates leaves a range from $0.584 million to

Table C-1. Estimates of Willingness-to-Pay to Avoid an Increased Risk of Fatality Implied by Wage Differentials (all dollars in thousands)

Authors	Published value	U.S. dollars of year	In 1983 U.S. dollars[a]
Thaler and Rosen (1975, p. 294)	140–260	1967	435–809
Brown (1980, p. 129)	400	1967	1,244
Smith (1976, p. 93)	1,500	1973	3,240
Viscusi (1979, p. 249)	900–1,500	1969	2,493–4,155
Olson (1981, p. 173)	3,200–3,400	1973	6,912–7,344
Arnould and Nichols (1983, p. 338)	223–289	1970	584–757
Viscusi (1981, p. 289)	3,200–4,160	1978	4,640–6,032
Value per life-year			
Thaler and Rosen (1975)[b]	8.4–15.7		26–49
Brown (1980)[c]	20.4		63

[a] U.S. dollars adjusted to 1983 using the following inflators: 1967, 3.11; 1969, 2.77; 1970, 2.62; 1973, 2.16; 1978, 1.45. 1983 U.S. dollars may be converted to Canadian dollars at 1.2324.
[b] Assuming an average of 32 years lost.
[c] Assuming an average of 52 years lost.

$6 million 1983 U.S. dollars, or from three quarters of a million to 7.4 million 1983 Canadian dollars.

The average age of the workers varied from about 42 years in the Thaler and Rosen study to about 20 years in the Brown study. Thus, assuming a life expectancy of 72 years, on average, about 32 years of life were lost for each accidental death in Thaler and Rosen, and 52 years in Brown. The economic model presented in chapter 5 values the discounted sum of life-years saved. Thus if the compensating differential reveals an average value of life of V, then according to that model, this must satisfy $V = U(x) \Sigma_{t+1}^{T} D^t$, where T is the number of years of life remaining, on average. The value of a life-year, $U(x_t)$, is then equal to $V \Sigma_{t+1}^{T} D^t$. Using a 4.6 percent discount rate (see chapter 5), the discount factor for 32 years is 16.58. The Thaler and Rosen study thus implies a value of $26,000 to $49,000 per life-year. The discount factor for 52 years is 19.64. The Brown study thus implies a value of $63,000 per life-year. Since the recent studies by Olson and Viscusi reveal values of life eight to ten times that of Brown, an upper estimate of the value of a life-year might be $600,000.

These market-derived values understate the average building occupant's value of life for two reasons. First, workers who choose particularly risky occupations presumably place a lower value on avoiding injury or death than does the average worker. Viscusi (1983) found that the compensating wage for workers in high-risk jobs was significantly less per unit of risk than that for workers in low-risk jobs. Presumably building occupants in general have less taste for risk than do industrial workers in occupations distinguished by their high risks. Second, workers who are injured receive workers' compensation benefits in addition to the compensating wage differential. Thus the total compensation for facing the risk is the compensating wage plus the expected value of future workers' compensation payments. The data in table C-1 report only the former, and therefore understate the value that workers place on avoiding mortality risks.

Having mentioned workers' compensation, it should be pointed out that workers' compensation payments for accidental death on the job are not properly interpreted as a measure of willingness to pay to reduce mortality risks. The workers' compensation system was designed to make up part, but not all, of the financial loss suffered by survivors and dependents of the injured worker. Death benefits are generally computed in relation to the deceased's wage rate and the number of survivors. In the case of a young worker with no survivors and no dependents, no payment might be made. There is no theoretical reason to believe that such payments would be related in any way to the amount that the deceased would have been willing to pay to reduce the risk of death. Similarly, tort awards for wrongful deaths are generally computed from discounted future earnings, net of the estimated consumption of the deceased. Again, these payments bear no necessary relationship to the willingness of the individual, or society on behalf of the individual, to expend resources to reduce the probability of fatal accidents.

Costs of Public Programs

The cost of public programs designed to reduce chances of premature mortality may be examined to determine what society is prepared to pay to prevent the loss of life in similar situations. If public decisions were made in a framework of economic rationality, and if the objective were to minimize the loss of life within some budget constraint, then we should be willing to spend the same amount at the margin for a given amount of life-saving. Viscusi (1981, table 1) and Graham and Vaupel (1981, p. 94) present some data on the implicit cost of public programs. These data are reproduced in table C-2, including a conversion to 1983 U.S. dollars. The costs range from $67,000 for consumer product safety to $83 million for regulating occupational exposure to arsenic. It appears that expenditures in the

Table C-2. Estimates of the Cost of Protecting Life: Public Programs ($ thousand)

	Expenditure per life saved	In 1983 U.S. dollars
Viscusi[a]		
Occupational Health Standards		
Acrylonitrite 2 ppm	4,600	5,574
Arsenic 0.004 mg/m³ (average)	5,600	6,786
Arsenic 0.004 mg/m³ (marginal)	68,100	82,522
Coke Oven 0.3 mg/m³	13,900	16,844
Graham and Vaupel[b]		
Median cost of many programs		
Traffic Safety	64	84
Health Services	102	135
Consumer Products	50	67
Environment	2,600	3,424
Occupational Health & Safety	12,100	15,932

[a] 1982, Table 1. The cost presented is the average cost per life at this standard, which is less than the marginal cost per life of this standard versus a less strict one. The arsenic standard of 0.004 mg/m³ causes a marginal cost of $68.1 million per life compared to 0.05 mg/m³ (Viscusi, 1982, p. 972). Original calculation in 1980 U.S. dollars.

[b] 1981, p. 94. Assuming original calculation in 1979 U.S. dollars. Conversion factor to 1983 U.S. dollars: 1.3145.

fields of occupational health and safety and environmental health are far more costly than those in other fields. Since the data used here do not, however, come from a scientifically representative study, it is unwise to generalize too broadly from them. These costs are up to ten times as great as the values derived from the compensating differential. We cannot tell whether this discrepancy reflects social values of life that exceed private values, measurement errors, or errors in public policy.

The studies above do not report the average age at death, and so it is not possible to calculate the implicit cost per life-year saved. However, if an average of twenty years of life were lost in each fatality, the value of a life-year would be about one-thirteenth the value of a life. By this calculation, only the arsenic standard substantially exceeds the level of $1 million per life-year.

REFERENCES

Arnould, Richard J. and Len M. Nichols. 1983. "Wage-Risk Premiums and Workers' Compensation: A Refinement of Estimates of Compensating Wage Differential," *Journal of Political Economy* vol. 91, no. 2, pp. 332–340.

Bailey, Martin J. 1980. *Reducing Risks to Life* (Washington, D.C., The American Enterprise Institute for Public Policy Research).

Brown, Charles. 1980. "Equalizing Differences in the Labour Market," *Quarterly Journal of Economics* vol. 94, (February) pp. 113–134.

Graham, J. D. and J. W. Vaupel. 1981. "The Value of a Life: What Difference Does It Make?" *Risk Analysis* vol. 1, no. 1 (March) pp. 89–95.

Rea, Samuel A., Jr. 1983. "Regulating Occupational Health and Safety," in Donald N. Dewees, ed., *The Regulation of Quality: Products, Services, Workplaces, the Environment* (Toronto, Butterworths).

Smith, Robert S. 1976. *The Occupational Safety and Health Act: Its Goals and Its Achievements* (Washington, D.C., The American Enterprise Institute for Public Policy Research).

Thaler, Richard and Sherwin Rosen. 1975. "The Value of Saving a Life: Evidence from the Labor Market," in Nestor Terleckyj, ed., *Household Production and Consumption* (New York, National Bureau of Economic Research) pp. 265–302.

Viscusi, W. Kip. 1979. *Employment Hazards—An Investigation of Market Performance* (Cambridge, Mass., Harvard University Press).

_____.1981. "Occupational Safety and Health Regulation: Its Impact and Policy Alternatives," in John P. Crecine, ed., *Research in Public Policy Analysis and Management*, vol. 2 (Greenwich, Conn., JAI Press) pp. 281–299.

_____.1982. "Setting Efficient Standards for Occupational Hazards," *Journal of Occupational Medicine* vol. 24, no. 10 (December) pp. 969–976.

_____.1983. "Alternative Approaches to Valuing the Health Impacts of Accidents: Liability Law and Prospective Evaluations," *Journal of Law and Contemporary Problems* vol. 46, no. 4, pp. 49–68.